Glossary of mushroom fruitbody terms

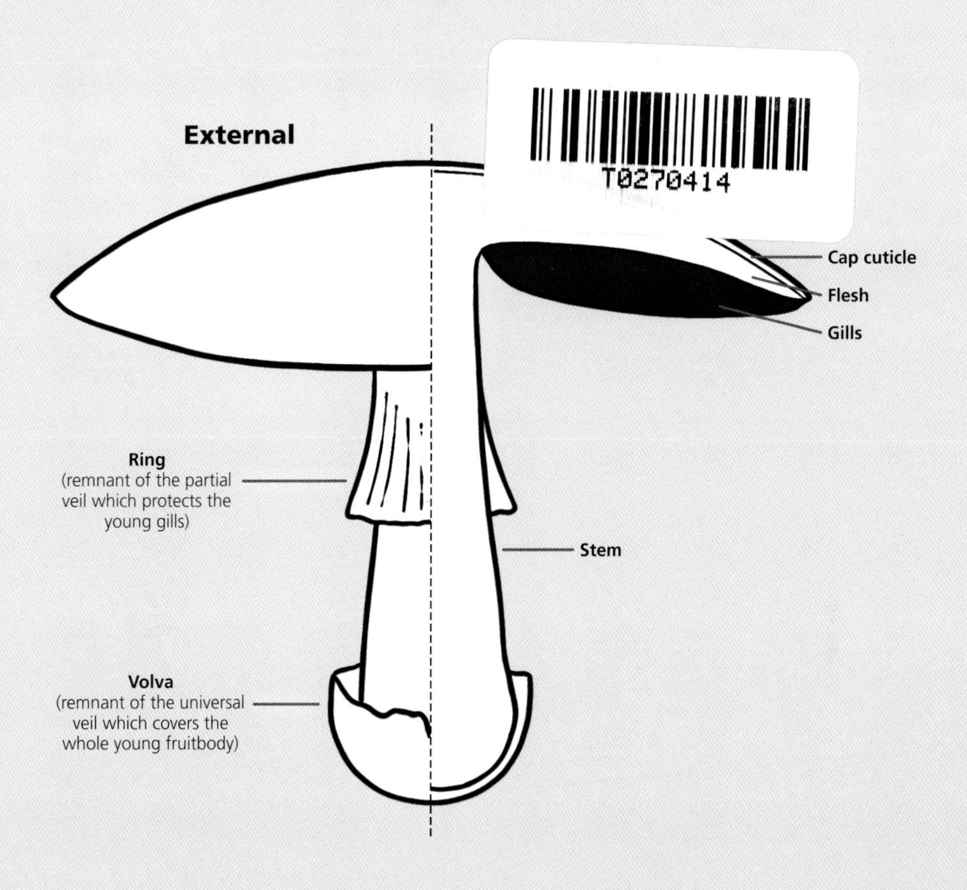

Gill attachments – see also page 114

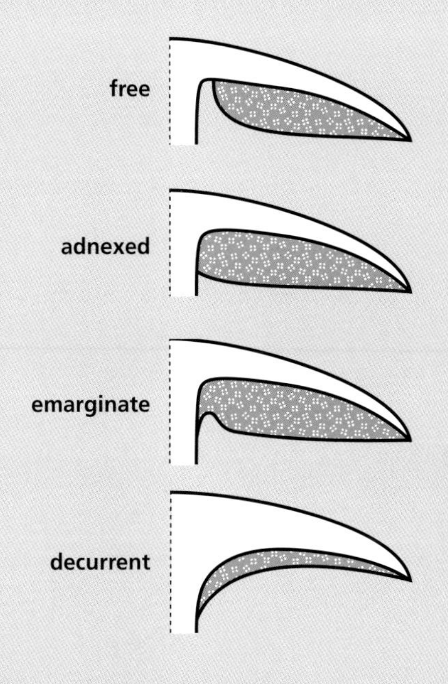

EDIBLE FUNGI of Britain and Northern Europe

How to Identify, Collect and Prepare

Jens H. Petersen

Princeton University Press
Princeton and Oxford

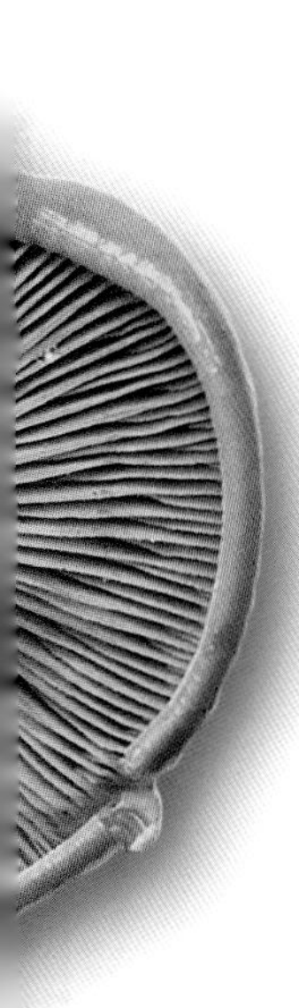

Wood Blewit (*Lepista nuda*)

This book is for those of you who want to collect and eat mushrooms for the first time. It is about the edible and easily recognizable fungi – and those that resemble them.

In Britain and Ireland, there are many thousands of fungal species that form large fruitbodies, and correct identification can be very hard. Precise species identification often requires much experience and the use of a microscope – and even then, it may be difficult.

You may be experienced in correctly identifying butterflies or plants, but you will not find it so easy with the fungal kingdom. Often the best that we can do is to narrow down the identity of a collected fungus to the correct major mushroom group, e.g. to the boletes or, slightly more precisely, to the cep group.

The central objective of this book is to enable you to recognize precisely these major groups of edible mushrooms, such as yellow chanterelles, blueing boletes with orange tube mouths, green brittlegills or milkcaps with orange milk. These species groups will usually be presented on a spread, where other similar or problematic fungi are also shown. Once you are familiar with some of these groups, your career as a mushroom chef can start.

Before the pages on mushroom groups you will find an introductory section: what fungi to look for, where to find them, how to collect and prepare them, etc. You may have been turned off by user manuals in the past, but this information is the key to mastering the skill of identification, so do yourself a favour by ploughing through these pages one dark and rainy autumn day, while waiting for the weather to improve. Then you can be ready too get out and look for edible mushrooms.

Good mushroom hunting!

carletina Bolete (*Neoboletus erythropus*)

Fungi are neither plants nor animals: they belong to their own kingdom, **the kingdom of Fungi**. They differ from plants by not being able to perform photosynthesis and from animals by being made up of long cell chains called hyphae.

The fungus itself – the fungal individual – consists of a branched network of hyphae that live in a nutrient substrate, for example in a layer of leaves or in a tree trunk.

Mushroom life

The life cycle starts with a microscopic **spore** that germinates and forms a thin, tubular cell called a **hypha**. If the hypha can find nutrients, it will grow and begin to branch. The resulting system of branched hyphae is called a **mycelium**. This will spread like a circle in search of more nutrients. We all know little fungal mycelia from the moulds on foods that have become too old.

If the fungus thrives and can find enough food, it may eventually form **fruitbodies**. These are the fleshy structures that we see in the woods on an autumn day. Within the fruitbodies sexual reproduction produces large numbers of spores which are released to disperse the fungus.

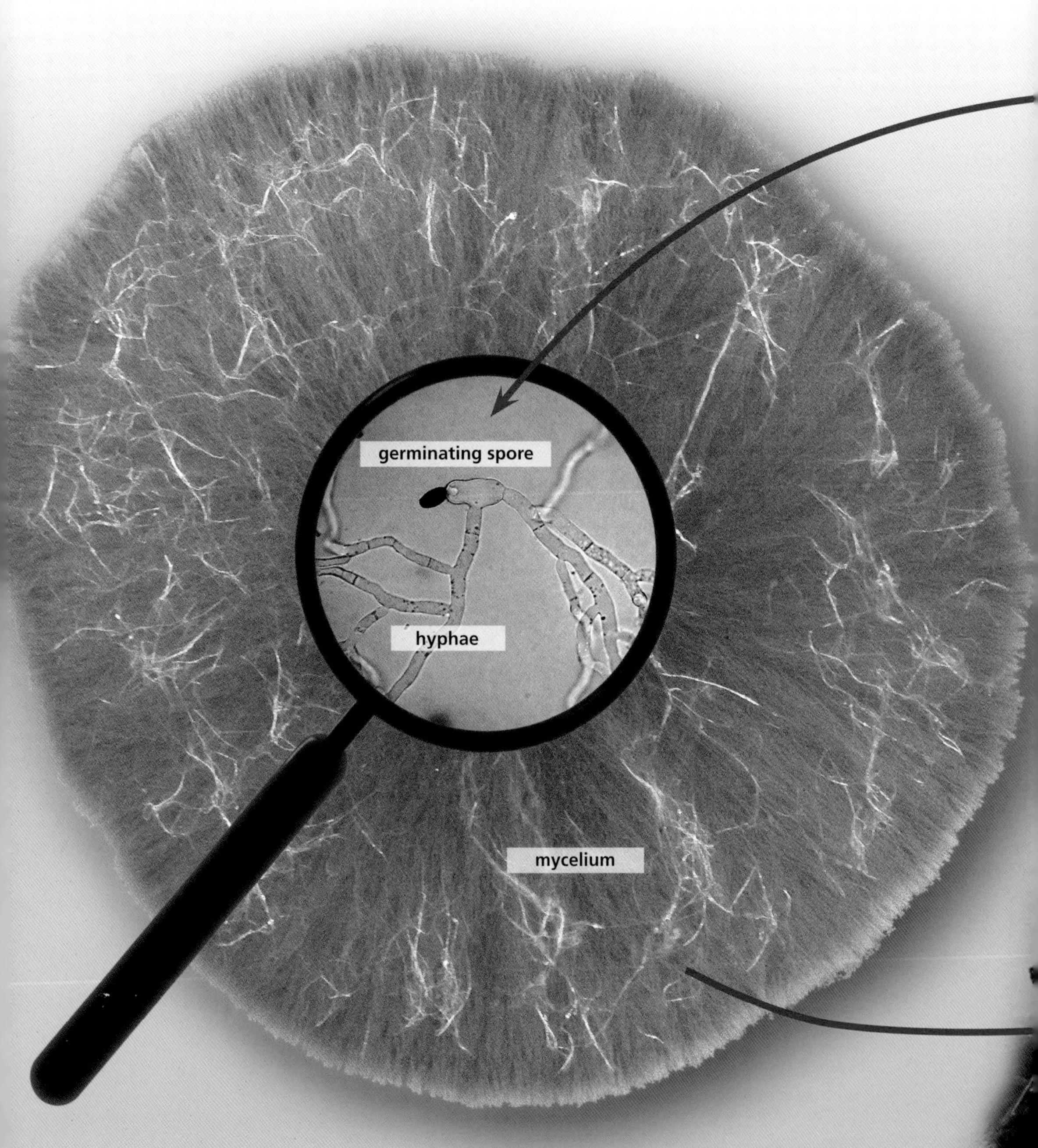

Superdiversity

The kingdom of Fungi is immensely species-rich. It is estimated that there are at least 1.5 million fungal species on the planet. Fungi are the second largest group of higher organisms, surpassed only by the insects of the animal kingdom.

In Britain and Ireland the number of known fungi is overwhelming, certainly more than 15,000 species, and as the exploration of fungal diversity is on-going, numerous new species are discovered every year, even in well-researched areas.

When it comes to collecting edible mushrooms in nature, there are not 15,000+ species to choose from. Many species do not form large fruitbodies or other eye-catching structures, and so remain invisible to us. Instead they live as mycelia within soil and organic matter and may only be revealed during DNA analyses or very laborious cultivation in the laboratory. But even if we subtract all the 'invisible' species, we still end up with several thousand fungi with fruitbodies large enough for us to notice on a mushroom hunt. Fortunately, the number of interesting edible mushrooms – the theme of this book – only reaches a few hundred species.

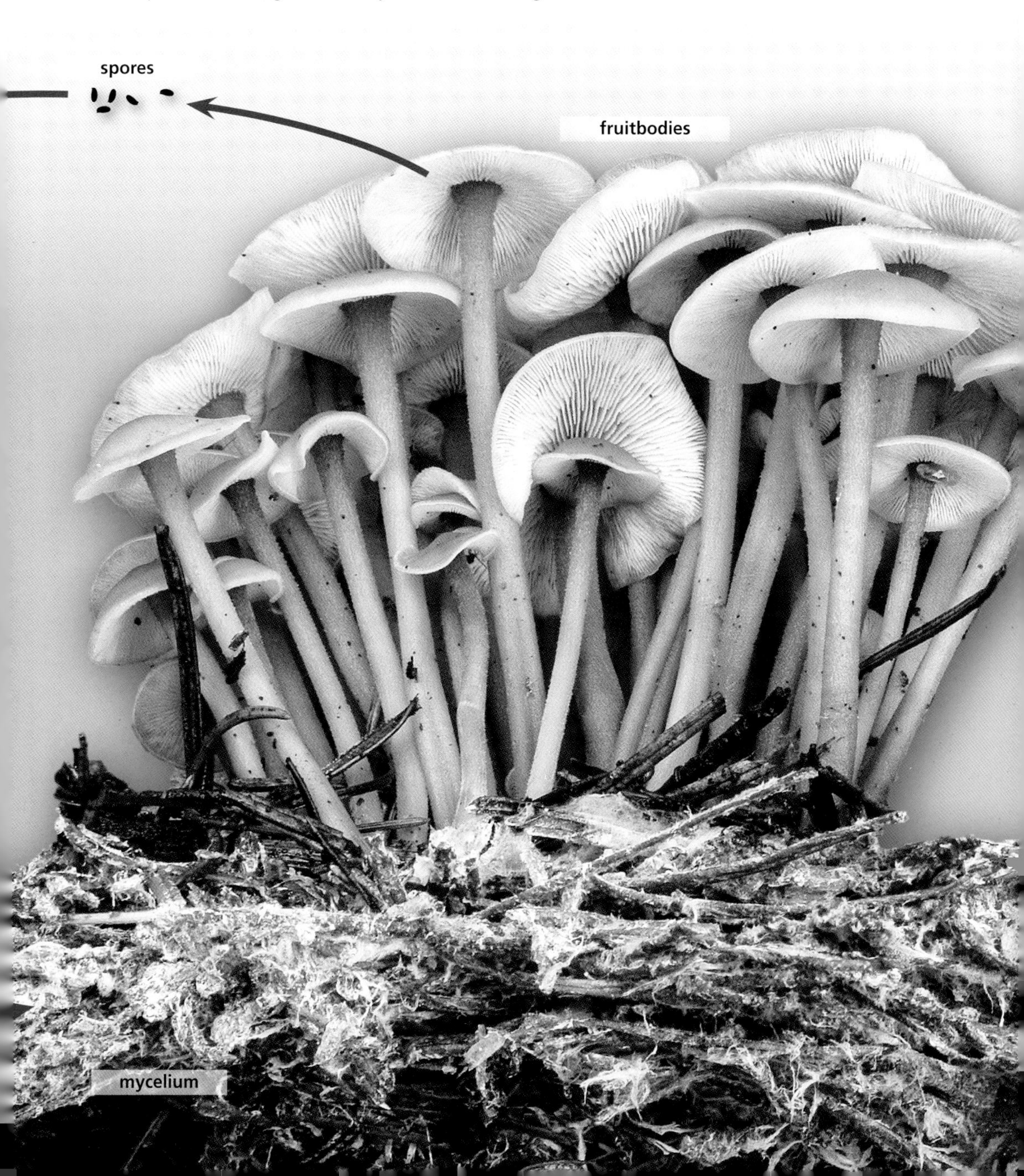

Since fungi cannot photosynthesize, they must, like animals, obtain their energy from the sugars generated by plants. This can happen in two different ways: through decomposition or by symbiosis (see page 12).

At the surface you only see the small fruitbodies, but beneath, a large mycelium spreads through the layer of branches and conifer needles

Decomposers

Among the ecological strategies of fungi we find some species that break down simple sugars. A mould growing on jam, for example, absorbs sugar molecules into its branched network of hyphae which provide the fungus with energy.

However, it is not just simple sugars that are metabolized by mushrooms. Most decomposer fungi have developed advanced enzyme systems that enable them to cope with very complex organic substances such as the cellulose and lignin found in the cell walls of plants. Thus fungi are the universal waste disposal system of nature. Without them, dead plants would simply accumulate; through decomposition the fungi ensure that dead organic matter is recycled.

A ram's head, where the horns are completely covered with fruiting bodies of the Horn Stalkball (*Onygena equina*)

Substrates

Decomposing fungi can thrive on every conceivable source of nutrition, from foods to living plants, plant debris, tree trunks, dung, horn and feathers, to exotic substrates like the coating of camera lenses, the emulsion of photographic film and aviation fuel. Each substrate requires a fungal specialist which can secrete exactly the right enzymes needed to dissolve it.

Fairy rings

To a decomposer fungus, a grassy field or the needle cover of a woodland floor may promise an endless source of food. The fungus can grow unobstructed in all directions, and since fungal mycelia naturally form circular patches, the ground will eventually be covered by large fungal circles. As they grow, the fungi use up the available nutrients, before expanding outwards. Therefore, while the mycelia may die away in the nutrient-depleted centre, the fungi will remain alive and growing at the outer edge of the circles, thus forming fairy rings.

Some fairy rings are noticed only when the fruitbodies appear in the autumn, whereas others can be seen from changes in the grassland itself. The growing mycelium can, for example, release excess nutrients so the vegetation is fertilized and appears greener than the surrounding grass. Alternatively the fungus may parasitize the grass, leaving a decayed, brown ring in the field.

Circular mycelia in a petri dish

A fairy ring formed by fruitbodies of Wood Blewit (*Lepista nuda*)

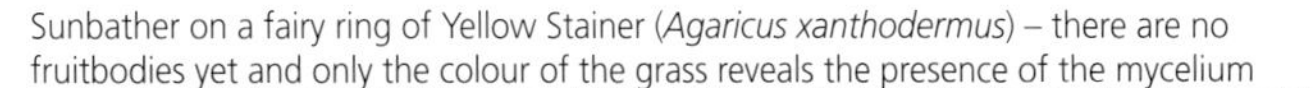

Sunbather on a fairy ring of Yellow Stainer (*Agaricus xanthodermus*) – there are no fruitbodies yet and only the colour of the grass reveals the presence of the mycelium

Wood decomposers

Decomposition of wood is a particularly challenging job. Wood contains large amounts of cellulose and lignin, and breaking it down requires an advanced enzyme system.

Several good edible fungi live as wood decomposers, such as Wood Cauliflower (*Sparassis crispa*), Chicken of the Woods (*Laetiporus sulphureus*) and Hen of the Woods (*Grifola frondosa*). However, there are also a number highly destructive fungi among the wood degraders. Jelly spots (*Dacrymyces*), mazegills (*Gloeophyllum*), Wet Rot (*Coniophora puteana*) and Dry Rot (*Serpula lacrymans*), for example, do damage by decomposing damp wood in houses.

Beefsteak Fungus (*Fistulina hepatica*) and Hen of the Woods at the foot of an old oak that may easily survive for decades despite the activity of the fungi

Large quantities of dead wood should be left in woodland for fungi to thrive

Parasites

Some wood decomposers do not just attack the wood of weakened or dead trees – they also attack living trees and kill them. In that way, they are able to prepare their own food and to be ahead of competing species.

Parasites like Root Rot (*Heterobasidion*) cost the forestry industry millions annually, and the edible Honey Fungus (*Armillaria*) also come high on the list of expensive forest pests.

An old beech stump being decomposed by honey fungi, brittlestems (*Psathyrella*) and many other species

Fungi in symbiosis

Mutual symbiosis is a coexistence of several living organisms to the benefit of all participants. Many fungi are involved in symbiosis with photosynthesizing plants. This allows the fungi to have direct access to sugars produced in the plants by photosynthesis.

Fungus–plant symbiosis can be divided into two types: **lichen formation** and **mycorrhizal**. In lichens the fungi form flat or branched structures which house single-celled green algae. The algae produce sugars for the fungi, the fungi provide protection for the algae and both symbionts thrive. They are completely integrated into a common 'individual' e.g. a Reindeer Lichen (*Cladonia*).

Mycorrhiza

A mycorrhizal association is also a coexistence between a fungus and a plant, but the two partners still appear as separate organisms.

In a mycorrhizal symbiosis, fungal mycelia grow around and into plant roots. In this way, the plants are able to extend their root system through the extremely thin fungal hyphae, forming a much finer network than that of the plant roots alone. The very large surface of the mycelium is highly efficient at absorbing water and nutrients which are then transported back to the plants.

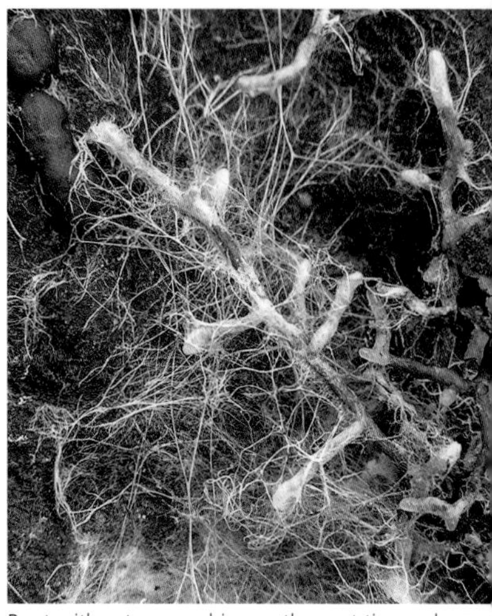

Root with ectomycorrhiza on the root tips and mycelium radiating into the surrounding soil

Ectomycorrhiza connect a spruce and a beech in a complex network where minerals and nutrients can flow between all parties involved

Common Chanterelle (*Cantharellus cibarius*)

False Saffron Milkcap (*Lactarius deterrimus*)

The Gypsy (*Cortinarius caperatus*)

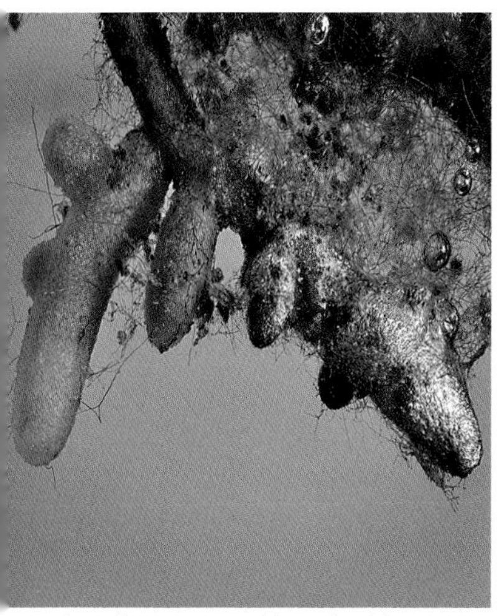

A small root system with a yellowish, a reddish and a greenish mycorrhizal fungus

In return, the fungus receives sugars from photosynthesis in the plant.

Both herbs and trees can form mycorrhizal associations. The type that involves edible fungi such as boletes (*Boletus* and others), brittlegills (*Russula*), milkcaps (*Lactarius*) and chanterelles (*Cantharellus*) is called ectomycorrhiza. Here the plant partner is almost always a tree. Typical ectomycorrhizal trees are beech, oak, lime, birch, alder, hazel, spruce, pine and fir. The characteristically branched mycorrhizal roots are located near the soil surface and can easily be excavated and studied.

The fungi and trees linked together by ectomycorrhiza are completely interdependent. If the fungi are missing, the woodland will not thrive and the fungi, for their part, will not be able to grow and form fruitbodies without their trees.

Some fungi can form mycorrhizal associations only with certain trees, while others, such as many of the amanitas, grow with several different tree species. Since a single tree usually has many different fungal partners, the mycorrhiza will connect both trees and fungi in a complex network. Nutrients can migrate freely through this network, so that sugars can flow from large trees to smaller ones, and even between different species of trees and fungi.

Greencracked Brittlegill (*Russula virescens*)

Destroying Angel (*Amanita virosa*)

Cep (*Boletus edulis*)

Wood Hedgehog (*Hydnum repandum*)

Horn of Plenty (*Craterellus cornucopioides*)

The open landscape – a mosaic of cultivated fields and grassland

Your first question is probably "Where do I find the mushrooms?" And the answer "Everywhere there is organic material" is both true but unhelpful. Let me instead try to divide nature into four categories:

- the open landscape
- parks, churchyards and cemeteries
- coniferous forests
- deciduous woodlands

Before setting off, pay attention to the rules of access to the countryside. In Scotland, there is a Right to Roam on almost all land, apart from the immediate surroundings of private dwellings. In England and Wales, there are open access rights over some 8% of the land, mostly in the uplands, as mapped by the Ordnance Survey, together with an extensive network of Public Rights of Way. However, in Ireland, north and south of the border, there are few public footpaths and general access is restricted to National Parks. Rights of access for recreational purposes rarely include, and sometimes specifically exclude, the right to collect mushrooms or other wild foods, although these activities are widely tolerated.

Giant Puffball (*Calvatia gigantea*) on a heavily fertilized, grazed field

The open landscape

This mainly consists of fields and meadows. The majority of fields are heavily fertilized, either through artificial or organic fertilizers. In addition, rainfall may contain of nitrogen evaporated from slurry.

Edible mushrooms in the open landscape are mostly found in grazed fields. Here you may find blewits (*Lepista*), funnels (*Clitocybe* and *Aspropaxillus*), parasols (*Macrolepiota* and *Chlorophyllum*), *Agaricus* species and puffballs (*Lycoperdon* and *Calvatia*). All of these may form impressive fairy rings that can be seen from a long distance, and even on aerial photographs.

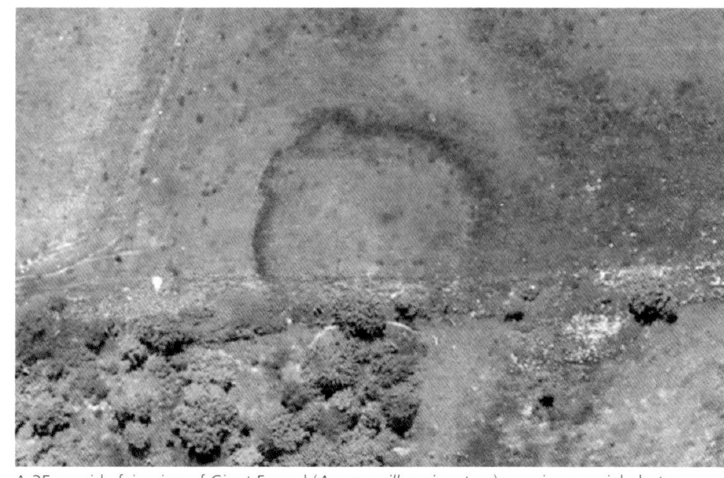

A 35 m-wide fairy ring of Giant Funnel (*Aspropaxillus giganteus*) seen in an aerial photo

On steep slopes which are difficult to plough, one may find grassland that for long periods has been grazed extensively by cattle, horses or sheep, and which has not received artificial fertilizer. The vegetation is often short with many different herbs and grasses, and there may also be scattered thorny shrubs like juniper and hawthorn. The fungal flora of ancient grassland can be very diverse, with many species of beautiful waxcaps (*Hygrocybe* and others), club fungi (*Clavaria* and others) and pinkgills (*Entoloma*) – but unfortunately not with many edible mushrooms. A single application of artificial fertilizer is enough to destroy the entire grassland flora and fungi. Then the areas are invaded by sorrels, nettles, thistles and cock's-foot – together with species of *Agaricus*.

A steep grassland with many beautiful waxcaps is a biological oasis

Both edible *Agaricus* species and here the slightly poisonous Yellow Stainer (*Agaricus xanthodermus*) are fond of damp lawns

Parks, churchyards and cemeteries
For those who live in towns and cities, the obvious places to start mushroom hunting are parks, churchyards and cemeteries. Here there will be both extensive lawns with decomposer fungi and trees such as oak, beech, birch, lime and conifers for the mycorrhizal species.

As a rule, there is widespread tolerance of mushroom collection in public spaces as long as appropriate consideration is given to other users. But if in doubt, you should always seek permission from the land owner or manager. And when collecting anything for consumption, you should always have a mind to the fact that dogs will do what dogs do!

Surrounded by the Blusher (*Amanita rubescens*)

There are many species of edible mushrooms that grow in parklands, though you may also find the slightly poisonous Yellow Stainer (*Agaricus xanthodermus*). The Fairy Ring Champignon (*Marasmius oreades*) often forms fairy rings all over the grass.

Among the mycorrhizal fungi, boletes, milkcaps (*Lactarius*), brittlegills (*Russula*) and amanitas are common. During a humid period in late summer, the delicious Lurid Bolete (*Suillellus luridus*) may be quite frequent on clay soil, but most often species of lower culinary interest like the Brown Birch Bolete (*Leccinum scabrum*) will dominate. Under the birches you may also find edible brittlegills such as Green Brittlegill (*Russula aeruginea*), but beware of the Deathcap (*Amanita phalloides*) which may well grow in similar places.

Hen of the Woods on an old oak may live for decades without killing the tree

Historic parklands offer special opportunities. Here there will be old trees and often also lying, dead trunks and limbs, and, with a little luck, you will find wood-inhabiting edible fungi such as Beefsteak Fungus (*Fistulina hepatica*) and Hen of the Woods (*Grifola frondosa*) on oak, Giant Polypore (*Meripilus giganteus*) on beech and Chicken of the Woods (*Laetiporus sulphureus*) on various deciduous trees, including oak.

Old park with large, dead trees and limbs – perfect for Chicken of the Woods and Hen of the Woods

Coniferous forests

All our common conifers (spruce, fir, pine and larch) form ectomycorrhizal associations, so coniferous forest can offer both mycorrhizal and decomposing fungi.

Among the mycorrhizal species, one can find various chanterelles (*Cantharellus*), Crab and Hintapink Brittlegills (*Russula xerampelina* and *Russula paludosa*), the orange-milked milkcaps (e.g. *Lactarius deterrimus*) and many boletes. The orange-milked milkcaps often grow along the tracksides, while the other species may be found scattered through the forest. If you travel to Scandinavia, you can also go hunting for Forest Lamb (*Albatrellus ovinus*) and The Gypsy (*Cortinarius caperatus*).

One of our very best edible mushrooms, the Common Chanterelle (*Cantharellus cibarius*), can be found in pine plantations as early as July, while Cep (*Boletus edulis*) can be found in large quantities during the 'Cep boom', for a few weeks around the 1st September.

The edge of a path through a coniferous forest, where milkcaps or chanterelles may grow

Nutrient-poor coniferous forest with reindeer lichens (*Cladonia*) and The Gypsy

Conifer plantations may offer numerous edible decomposing fungi during late autumn. On the roots and stumps there may be large amounts of Dark Honey Fungus (*Armillaria ostoyae* – however, see the note on its digestibility on page 122), and the needle layer can abound with fairy rings of Wood Blewit (*Lepista nuda*). Delicious species of *Agaricus* may also be frequent, but beware of the deadly poisonous Destroying Angel (*Amanita virosa*) that grows in similar places.

Finally, you may be lucky enough to find a coniferous tree with Wood Cauliflower (*Sparassis crispa*) at the base. This will return every autumn for a number of years, until the wood is eaten up: a couple of secret Cauliflower stumps would sit well in your fungal portfolio.

Spruce plantation with a mat of brown needle cover – a good place to look for Cep and Wood Blewit

Dead conifer wood may produce honey fungi and Wood Cauliflower

Beech woodland on a clayey coastal slope – a great place to look for brittlegills and chanterelles

Deciduous woodlands

Deciduous woodlands are good places to discover edible fungi, both mycorrhizal and wood-inhabiting.

Fungi can occur anywhere in a woodland, but you may have the best luck looking along roadsides and tracks or on slopes. In these places road material and sliding gravel create mixed soil types resulting in a large number of microhabitats with differing nutritional conditions and pH levels.

Among the ectomycorrhizal fungi you will have a good chance of finding delicious edible species like boletes, brittlegills (*Russula*), hedgehog fungi (*Hydnum*), chanterelles (*Cantharellus*) and Horn of Plenty (*Craterellus cornucopioides*). But beware: you may find the deadly poisonous Deathcap (*Amanita phalloides*) and Destroying Angel (*Amanita virosa*) in the same places.

To hunt for ectomycorrhizal fungi, you must look for trees like beech, oak, hornbeam, lime, hazel, willow, poplar and birch. Ash and maple, on the other hand, do not form ectomycorrhizal associations.

Among the common edible wood-inhabiting fungi are species of Honey

A slope with exposed soil comprising both acidic and alkaline areas, where Deathcap has found a place to grow

A beech woodland where the dead trunks have been allowed to remain – to the benefit of both edible fungi and Tinder Bracket

Fungus (*Armillaria*). They mostly fruit from roots and stumps in mid-autumn. Later in the year you can hunt for Oyster Mushroom (*Pleurotus ostreatus*) on larger trunks of, for example, beech, willow and poplar. During summer and autumn look for Chicken of the Woods (*Laetiporus sulphureus*) and Hen of the Woods (*Grifola frondosa*) on the old oaks. You may also enjoy the sight of the perennial Tinder Bracket (*Fomes fomentarius*), which is common on beech and birch, but is as hard as wood, and can't be eaten!

Old, grazed oak wood – a good place to find Chicken of the Woods and Hen of the Woods

Edible fungi can be found almost all year round – but most species form fruitbodies during autumn.

Spring offers the rare morels (*Morchella*) and the more common St George's Mushroom (*Calocybe gambosa*). Early summer brings the first chanterelles (*Cantharellus*), brittlegills (*Russula*) and boletes and autumn offers a wide range of edible mushrooms (see the chart below).

Most edible fungi form fruitbodies only when their environment is moist. Often there is a dry period with few visible fungi during summer, but usually about four weeks after the first autumnal rains the fruitbodies appear – first the decomposers, followed by the mycorrhizals.

During prolonged drought, one can successfully search for polypores on large trunks. The rotten wood retains water much longer than the soil, thus polypores are more able to withstand drought than many other fungi.

Edible fungi can be found throughout the autumn, right up until frost destroys the fruitbodies. During winter, only a few cold-adapted species such as Oyster Mushroom (*Pleurotus ostreatus*), Velvet Shank (*Flammulina velutipes*) and Jelly Ear (*Auricularia auricula-judae*) remain.

When to find edible fungi

	Jan	Feb	Mar	Apr	May	Jun	Jul	Aug	Sep	Oct	Nov	Dec
Black Morel	·	·	·	••	••	·	·	·	·	·	·	·
Common Morel	·	·	·	•	••	·	·	·	·	·	·	·
St George's Mushroom	·	·	·	·	••	••	·	·	·	·	·	·
Dryad's Saddle	·	·	·	·	••	••	•	•	•	·	·	·
Chicken of the Woods	·	·	·	·	•	••	•	••	••	·	·	·
yellow chanterelles	·	·	·	·	·	•	••	••	••	•	·	·
species of *Agaricus*	·	·	·	·	·	•	•	••	••	•	·	·
red-tubed boletes	·	·	·	·	·	·	••	••	••	•	·	·
Cep	·	·	·	·	·	•	••	••	•	·	·	·
yellow-tubed boletes	·	·	·	·	·	·	•	•	••	•	·	·
slimy boletes (*Suillus*)	·	·	·	·	·	·	•	•	••	••	·	·
brittlegills	·	·	·	·	·	·	•	••	••	•	·	·
Fairy Ring Champignon	·	·	·	·	·	·	•	••	••	•	·	·
Giant Puffball	·	·	·	·	·	·	·	••	•	•	·	·
Giant Funnel	·	·	·	·	·	·	·	••	••	·	·	·
Forest Lamb	·	·	·	·	·	·	·	••	••	·	·	·
scaly-stemmed boletes (*Leccinum*)			·	·	·	·	·	••	••	•	·	·
Giant Parasol	·	·	·	·	·	·	·	•	••	••	·	·
puffballs	·	·	·	·	·	·	·	•	••	••	·	·
Giant Polypore	·	·	·	·	·	·	·	•	••	•	·	·
Beefsteak Fungus	·	·	·	·	·	·	·	•	••	•	·	·
Slimy Spike	·	·	·	·	·	·	·	•	••	•	·	·
The Gypsy	·	·	·	·	·	·	·	•	••	•	·	·
hedgehog fungi	·	·	·	·	·	·	·	•	••	••	·	·
Horn of Plenty	·	·	·	·	·	·	·	•	••	••	·	·
milkcaps	·	·	·	·	·	·	·	•	••	••	·	·
Garlic Parachute	·	·	·	·	·	·	·	•	••	••	•	·
Trumpet Chanterelle	·	·	·	·	·	·	·	•	••	••	•	•
Shaggy Inkcap	·	·	·	·	·	·	·	·	••	••	•	·
Umbrella Polypore	·	·	·	·	·	·	·	·	••	••	·	·
honey fungi	·	·	·	·	·	·	·	·	•	••	•	·
blewits	·	·	·	·	·	·	·	·	•	••	••	•
Oyster Mushroom	••	•	•	·	·	·	·	·	·	••	••	••
Velvet Shank	••	•	•	·	·	·	·	·	·	·	•	••
Jelly Ear	••	•	•	•	•	•	•	•	•	••	••	••

Wood Blewit (*Lepista nuda*) in sn

Edible mushrooms

This is a book about the best edible mushrooms. Being a good edible mushroom is a combination of being tasty, easily recognizable and fairly common.

Traditionally, mushrooms have not been popular on the menus of Western European countries. They were often considered dangerous or even associated with the supernatural and evil. Those who collected fungi were suspect or eccentric. Only during the 20th century has the use of wild fungi become more acceptable.

If you ask Western European mushroom collectors for a list of their top fungi, a rather uniform picture emerges: the preferred edible mushrooms are the yellow chanterelles, the boletes in the cep group, the marzipan-smelling *Agaricus* species and morels. In this book the following have been given the top grade of two green dots (ordered as they occur in the book):

Common Morel (*Morchella esculenta*)
Black Morel (*Morchella elata*)
Wood Cauliflower (*Sparassis crispa*)
Hen of the Woods (*Grifola frondosa*)
Forest Lamb (*Albatrellus ovinus*)
the ceps (*Boletus* species)
red-tubed boletes (*Suillellus/Neoboletus* spp.)
the hedgehogs (*Hydnum* species)
yellow chanterelles (*Cantharellus cibarius* group)
Horn of Plenty (*Craterellus cornucopioides*)
Crab Brittlegill (*Russula xerampelina*)
Fishy Milkcap (*Lactifluus volemus*)
Saffron Milkcap (*Lactarius deliciosus*)
the royals (the marzipan-smelling *Agaricus* spp.)
The Gypsy (*Cortinarius caperatus*)

Fungal nutrients

In terms of nutrition mushrooms have quite a lot to offer. They do not contain much fat and sugar, but are rich in protein.

Since the fungal kingdom is more closely related to animals than to plants, it is not surprising that some amino acids which are difficult to get from plant diets can be obtained through fungi. Also Vitamin B12, which is usually obtained from animal products, is found in a fairly large amount in some fungi. On the whole, mushrooms contain a lot of vitamins and provide a useful supplement to a vegetarian diet.

Traditionally, fungi have been only a small component of typical Western European and North American food. In other parts of the world the tradition is different: in Eastern Europe wild mushrooms have always been valued, in Africa wild mushrooms are in many places an important source of trade and nutrition and in Asia they not only use lots of wild mushrooms, but have also for centuries cultivated mushrooms for eating and for use in traditional medicine.

Cultivated Shiitake

The fungus eater's guidelines:

- **Only eat mushrooms when you feel absolutely sure of their identity.**
- **Only use young, fresh fruitbodies.**
- **Only eat raw mushrooms when it is explicitly stated that the species can be eaten raw: many species are indigestible or slightly poisonous without cooking.**
- **Always start out carefully with new edible species: you could be allergic.**

In recent years, quite a lot of research has gone into the medicinal properties of fungi. Since many antibiotics (e.g. penicillin) come from micro-fungi, interest has naturally been greatest in this area, but some of the edible fungi prized in Asia are also under scrutiny. Hen of the Woods (*Grifola frondosa*, see page 60) and the cultivated Shiitake (*Lentinula edodes*) contain, for example, substances thought to activate the immune system and possibly also to have a certain anti-cancer effect. Similar properties have been claimed in connection to many other fungi, e.g. the polypore Turkeytail (*Trametes versicolor* – or Kawaratake in Japanese), species close to Lacquered Bracket (*Ganoderma lucidum* – Reishi) and the Jelly Ear (*Auricularia auricula-judae*, see page 150).

The symbols used in this book

•• excellent **after proper cooking**
• acceptable to eat **after proper cooking**
• worthless
† poisonous
†† deadly poisonous

edible after proper cooking

worthless

poisonous

Poisonous fungi

A book on edible fungi is not complete if it does not include the main poisonous fungi that can be confused with the edible ones. Among the deadly species included are:

False Morel (*Gyromitra esculenta*), p. 53
Deathcap (*Amanita phalloides*), p. 129
Destroying Angel (*Amanita virosa*), p. 130

These three species are extensively discussed and illustrated, and they should easily be distinguished from the book's edible fungi.

In Northern European mushroom literature, four additional deadly species often appear:

Funeral Bell (*Galerina marginata*)
Splendid Webcap (*Cortinarius splendens*)
Deadly Webcap (*Cortinarius rubellus*)
Fool's Webcap (*Cortinarius orellanus*)

These brown-spored fungi, shown at the bottom of the opposite page, are not described further because they do not really resemble any of the edible fungi included in the book.

The red Fly Agaric is not as poisonous as its reputation suggests – you become sick, but healthy people will not die from eating it

Lots of less poisonous, yet unpleasant, species are mentioned in the book, such as Devil's Bolete (*Rubroboeltus satanas*), Bilious Bolete (*Rubroboletus legaliae*), Fly Agaric (*Amanita muscaria*), Panthercap (*Amanita pantherina*), Common Inkcap (*Coprinosis atramentarius*) and Livid Pinkgill (*Entoloma sinuatum*). These can all cause quite unpleasant poisonings, but none of these are usually life-threatening.

Toxins

The fungal toxins can be divided into four categories: gastrointestinal irritants, neurotoxins, cytotoxins and other toxins.

The **gastrointestinal irritants** provide the most harmless of the poisonings: rumbling, pain, diarrhoea and vomiting. The discomfort strikes quickly after eating the mushrooms and usually goes away again after a few hours. Different people often have different tolerances towards these

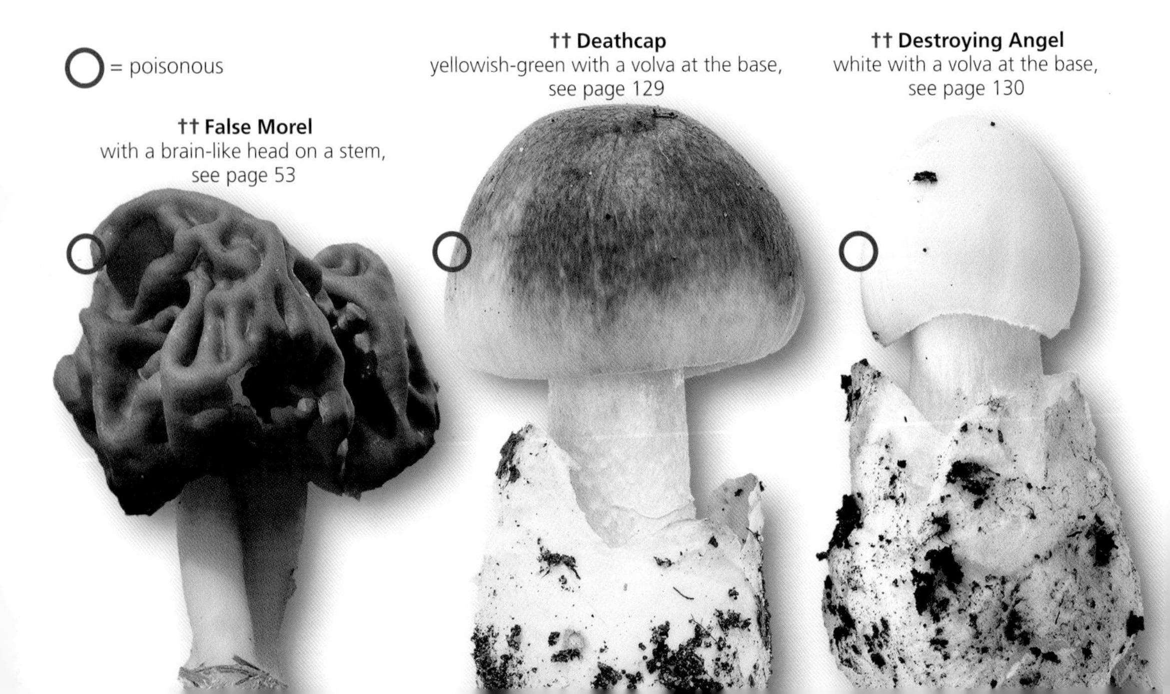

††Deathcap
– can be touched without
wearing plastic gloves!

mushrooms. For example, the Yellow Stainer (page 139) and the orange-capped boletes in the genus *Leccinum* (page 83) may cause a stomach upset in some while others can eat them after proper cooking without any problems.

The effects of **neurotoxins** will usually also be felt soon after ingestion. Symptoms may include dizziness, headache, confusion and hallucinations. As a rule, the poisonous effects disappear rather quickly. Fly Agaric (page 131), for example, contains neurotoxins, as does the strongly hallucinogenic Liberty Cap (*Psilocybe semilanceata*).

The really dangerous poisonous fungi usually contain **cytotoxins**. The poisonings typically are not felt until many hours, or even days, after the mushrooms have been eaten. Therefore, stomach pumping does not help and there are no effective antidotes. Typically, the toxins destroy the cells in the liver and kidneys. It is cytotoxins that make Deathcap and Destroying Angel such dangerously poisonous fungi. All of the deadly poisonous species listed at the bottom of this spread contain cytotoxins.

Under the category '**other toxins**' we find a number of more weird toxins, such as the relatively harmless Antabuse-like substance in Common Inkcap (page 141) and the more serious antigens in rollrims (*Paxillus*), see page 89.

Superstitions

If you listen to folk wisdom, a wide range of false advice abounds on how to differentiate edible fungi from poisonous ones. Mushrooms, for example, are claimed to be edible if they 'look delicious', 'smell good', 'taste good', 'are eaten by animals' or 'if a silver spoon does not darken when cooked with the mushrooms'. I even once heard a man and a woman in a pub discussing the use of a pendulum to divine whether or not mushrooms were 'good'. **None of this will work! Our most dangerous poisonous fungus, the Deathcap, looks quite delicious, smells and tastes mild, is eaten by larvae and snails, but on the other hand does not colour any silver spoons** (I do not know how it pendulates!).

Another superstition is that one should not touch poisonous mushrooms without gloves. If that were true, I would have long exhausted my nine lives during my career as a mushroom educator. So let it be clear: it is quite safe to touch the poisonous mushrooms – children should not wear plastic gloves when they go to the forest!

†† Deadly Webcap
orange-brown, brown spores
(**†† Fool's Webcap** looks the same)

†† Funeral Bell
small, with a ring,
brown spores; on wood

†† Splendid Webcap
yellow, with brown spores and
an edged bulb

Fungi play a key role in every terrestrial ecosystem, decaying and recycling the nutrients from dead organic matter, servicing plants with mycorrhizal and lichen connections and feeding a host of animals, especially invertebrates. It is therefore the responsibility of all who collect fungi for whatever purpose, whether for eating or identification, to do so in a sustainable manner, taking only what the environment and the species can withstand.

With this in mind, foragers must know their fungi, so that they can avoid picking, removing and potentially damaging unpalatable and threatened species. Please take only those specimens you know to be at a suitable stage for consumption: leave overmature ones for the beetles, fungus gnats and other insect larvae to eat. And try to avoid damaging nature by excessive trampling on sensitive wildlife habitats.

Bearded Tooth (*Hericium erinaceus*) grows on very old beech trees. It is on the British Red List and also protected by the Wildlife & Countryside Act 1981 and must not be collected.

The above advice should apply to all fungus collection. There are, however, additional circumstances which must be considered. Particular restraint should apply to any of those 400 scarcer species on the British national Red Data list (see the *Red Data List of Threatened British Fungi* on the British Mycological Society (BMS) website britmycolsoc.org.uk) or in the UK biodiversity action plans (see List of UK BAP Priority Fungi Species (including lichens) on the JNCC website, data.jncc.gov.uk).

Among the recommendations of the BMS Code of Conduct for Responsible Collecting of Fungi is that foraging should be actively discouraged on land designated for its wildlife conservation interest; the Society stresses its position that everyone should have the opportunity to enjoy the fleeting beauty of fungi undisturbed in the wild. There is as yet no Red List of fungi for the island of Ireland.

Furthermore, in Britain some species of fungi are specially protected by the Wildlife and Countryside Act 1981 (as amended) against picking and 'uprooting'. To take specimens, intentionally damage or destroy, or trade specimens of any of these would render the picker liable to a hefty fine. Four species are currently listed, none of which is covered in this book.

While mushroom hunting is normally accepted, it is in fact illegal to 'uproot' any 'wild plant' without the landowner's permission. For purposes of the law, plants include fungi, and uprooting is defined as 'dig up or otherwise remove the plant from the land on which it is growing', whether or not it actually has roots! Non-commercial collecting is generally tolerated and indeed the 1968 Theft Act specifically excludes the non-commercial picking of mushrooms from the definition of 'stealing'. However, some authorities and landowners (e.g. in some national parks and other public spaces) have passed bye-laws to reinforce a ban on the commercial collecting of fungi without a licence. Naturally all such regulations must be respected: they are in place for a purpose, in areas of high fungal diversity and importance and where overcollection has been a conservation issue.

The precise provisions for conservation vary to some extent between administrations within the UK, and also between Britain and the Republic of Ireland. The intending forager should ensure that they have an up-to-date understanding of policy and legality: ignorance of the law is no defence!

Conifer Roundhead (*Stropharia horneman*
only grows in the north (Scotland).
designated as Critically Endangered on
British Red List and should not be collect

Equipment

Before you go mushroom hunting, you need some equipment.

First and foremost, the fruitbodies must be picked and cleaned, so you need a knife. It can be a small kitchen knife or, as I prefer, a good pocket knife. You can also get special mushroom knives with a small brush that can be used to remove dirt.

If you want to collect agarics, it is good to know their spore colour. Here a 10x–20x hand lens is a great help.

You will also need something in which to transport the catch. There are many possibilities, as long as the fruitbodies are not squeezed. The classic is an open basket, but a flat-bottomed paper bag with a couple of grape trays or ice cream boxes inserted is also usable. Soft plastic bags should be avoided, as the fungi get mashed up and become unrecognizable. It also becomes very difficult to clean the fungi after returning home.

Quality

Only young, firm and healthy fruitbodies are suitable for cooking. Unfortunately, many species are attacked by insect larvae. There is actually a whole group of flies, the fungus gnats, whose larvae survive by eating fungal fruitbodies. Therefore, start by cutting a fruitbody to see what is going on inside.

Not only larvae pose a problem. Other fungi can also attack our edible fungi, especially moulds similar to those which grow on other foodstuffs. Boletes in particular are susceptible to moulds. So if the fruitbodies are strangely malformed, soft or with white or yellow fur, they must be left where they are.

Here stood a Horse Mushroom (*Agaricus arvensis*) – now only a slimy blob of larvae remains

The Matt Bolete (*Xerocomellus pruinatus*) on the right has been attacked by mould – you should leave mouldy fruitbodies where they grow

Picking techniques

You can pick the mushrooms with or without the basal part and soil.

If you do not know which fungus you have found, it is important to bring the whole fruitbody home. You should use your knife to gently lever the fruitbody from the ground, taking care to include the base as this may carry important characteristics for later identification.

If you are absolutely familiar with the species and want to collect for eating, you should instead clean the fruitbodies on the spot. You can then either cut the stem above ground or harvest the whole fruitbody and then remove the base with its dirt.

It has been a matter of debate which picking method is most benign for the fungus. Studies have, however, shown that they are equally good: the real fungus is the mycelium that lies within the soil or the tree trunk, and the mycelium is quite unaffected by the the way the fruitbodies are harvested.

Clean the edible fungi before they reach the basket

Preparation

Before you can start cooking, the mushrooms must be cleaned, and at this point you will reap the rewards of having cleaned the fungi well during the collecting process. Dirty stems that have sprinkled dirt on otherwise clean fruitbodies can destroy a whole basketful, and soil and sand between gills is almost impossible to get rid of.

Of course, it would be tempting to just wash the whole catch in water, but water dilutes the already very watery fruitbodies, so the golden rule is to avoid water as much as possible (depending on your tolerance of spruce needles and leaf residues in food). For fungal species that tend to become slimy during cooking, e.g. boletes, one should especially limit the use of water. Other species, such as Horn of Plenty and Hen of the Woods, are more tolerant of liquid.

The best way to clean mushrooms is to remove as much dirt as possible with a small knife or a brush and then wash with a little water where necessary. Finish by cutting the mushrooms into smaller pieces ready for cooking.

Cleaned Lurid Boletes – larger pieces need to be cut into smaller ones before cooking

The best edible mushrooms

There are thousands of mushroom species out there, but which ones are worth picking and how can they be cooked? In an attempt to rank the book's edible mushrooms, you may consider a combination of taste and consistency after cooking (see the chart below).

The species in the category 'good taste, firm consistency' are among the best edible mushrooms. They are easy to cook and have a characteristic taste of more than just 'mushroom'. Morels, chanterelles, Horn of Plenty, hedgehog fungi, Hen of the Woods and the good boletes and milkcaps all have their own

Good taste, firm consistency:	Good taste, soft consistency:	Mild taste, firm consistency:	Mild taste, soft consistency:
morels	Cep	many brittlegills	many boletes
Wood Cauliflower	St George's Mushroom	Chicken of the Woods	Oyster Mushroom
Hen of the Woods	honey fungi	Giant Polypore	reddening *Agaricus* species
Forest Lamb	marzipan-smelling *Agaricus* spp.		Shaggy Inkcap
red-tubed boletes	The Gypsy		many other agarics
chanterelles			
Horn of Plenty			
hedgehog fungi			
Crab Brittlegill			
Saffron Milkcap			
Fishy Milkcap			

Common Morel ready for cooking

quite indescribable tastes and must definitely be tried.

However, many of the other species can also provide excellent food, but here a little more cooking is often required to get a good result. For example, Cep is high on most people's wish list, but the art of frying Cep without just getting a sticky mass must be learned.

Species that do not taste of much, are most fun if you can mix them with the more flavourful species. This is true of all the species that are only awarded one green dot in this book.

Some species are outside these categories: blewits, for example, have a perfumed taste which some people like and others hate; Jelly Ear is strangely cartilaginous, but can be baked into pâtés and pies; Garlic Parachute may be used as an onion seasoning. Giant Parasol and Giant Puffball must be breaded to be worth the effort. Finally, there is the Beefsteak Fungus, which is soft and very sour, but is an unusual aesthetic experience when used raw in sushi.

Orange Birch Bolete (*Leccinum versipelle*) is a much-needed dietary supplement on a mountain hike with freeze-dried food

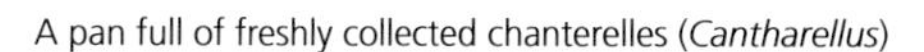

A pan full of freshly collected chanterelles (*Cantharellus*)

There are as many opinions about edible mushrooms and mushroom preparation as there are chefs. In this book, we try to keep it simple. Advanced mushroom gastronomy can be found in specialist mushroom cookbooks.

Fried and stewed mushrooms

The mother of all mushroom recipes is fried and stewed mushrooms. When you want to taste a new edible species, you should use this method so that you get the best impression of the basic taste of that particular fungus.

Frying fungi is easy: cut the cleaned mushrooms into centimetre-sized pieces and toss them in a hot pan (a non-stick pan is best). As a rule, they will start to release a lot of water, which should be allowed to boil away. When the pan becomes dry, add some olive oil (or butter) and continue frying until the pieces turn brown at the edges. Finally, add some cream and season with lemon and salt. Serve on a piece of toast or as an accompaniment to another dish.

The dish can of course be varied by frying the mushrooms with finely chopped onions or similar, but be careful not to mix with too many other flavours as they can easily overwhelm the mushroom taste.

Some species, e.g. Hen of the Woods, contain very little liquid, so with these you may use oil or butter from the start, but most fungi have to lose up to 80% of their weight in water before they can fry.

Boletes are especially tricky to get right because they easily become slimy. As a rule, you must be careful not to overfill the pan, and you must fry them hard and relentlessly to make them crisp. It is much easier to fry very young boletes, as even by 'middle age' they become problematic (with these you should remove the tubes). Older specimens are worthless. An interesting variation is to slice young Ceps very thin and fast-fry them in plenty of very hot oil. In this way, the fungi hardly shrink and they keep some of their nut-like raw taste.

Pasta with mushrooms

A variation is to stir plenty of cooked fungi with boiled pasta. All generations of my family love, for example, Wood Cauliflower, Hen of the Woods or Chicken of the Woods prepared in this way.

The classic: fried chanterelles with cream

Spinach pasta with Chicken of the Woods – a beautiful sight

Mushroom soup

A good soup is another simple mushroom dish: fry some mushrooms and chopped onions in a pot. Then add water and stock and bring it to the boil. Add cream, balance with lemon and salt and you have the nicest mushroom soup. If the mushrooms are a bit old or have a poor consistency, the soup can be blended.

There are lots of possible variations on this basic recipe, but again remember not to overwhelm the mushroom taste with too many other strong flavours.

Many species are suitable for soup, and if you have fungi smeared in sand, soup is the obvious choice, as the sand ends up at the bottom. I especially like a soup made from the yellowing, marzipan-scented *Agaricus* species.

Agaricus soup with chopped parsley

Pies and pizza

Mushrooms can also be used in pies or as pizza topping. Species that tend to soften during cooking work well in pies. Make some pastry by mixing 100 ml of water and 100 ml of olive oil with 300 g of wheat flour. Let it rest for a while in the fridge, then roll it thinly and line a couple of pie dishes. Pre-bake the bases at 200 °C until they are light brown. Fry the mushrooms until they have released their water and place them in the pie. Chop a few onions and cloves of garlic, and sprinkle them over the mushrooms. Stir four eggs together with some milk and pour over the filling. Bake at 200 °C until the surface is golden brown.

Horn of Plenty, Trumpet Chanterelle, boletes and Shaggy Inkcap are all very suitable fungi for pies, and the recipe can be varied by adding other vegetables too.

For pizza topping Horn of Plenty and Trumpet Chanterelle are magnificent. These contain very little water and can therefore be used raw together with, for example, onions, garlic, red pepper and cheese. They also taste strong enough to compete with the other ingredients.

Fresh Horn of Plenty

Pie with young Shaggy Inkcaps (*Coprinus comatus*)

Breaded mushrooms
Two of the species included in this book – Giant Puffball and Giant Parasol – should be coated in breadcrumbs before being cooked. Take hard fruitbodies of Giant Puffball, typically 10–20 cm in diameter, and cut them into centimetre-thick slices. With Giant Parasol use the 15–25 cm-wide flat caps without the stem.

Coat the mushroom in beaten egg and white breadcrumbs, season with salt and pepper, and fry the whole thing in the pan until it turns golden brown.

A slice of breaded Giant Puffball

Raw mushrooms in cooking
Because many fungi are indigestible or even poisonous before cooking, only a few species are recommended to be eaten raw. These include boletes of the cep group (but no other boletes), some species of *Agaricus* and the peculiar Beefsteak Fungus.

You can, for example, use Cep and most *Agaricus* species raw in salads. The fruitbodies must be very young and fresh, such as Cep in the champagne cork stage, see page 72.

Young Beefsteak Fungus has a very flesh-like inner structure. If you cut it thinly, it can become a very exotic element in sushi. The downside is the remarkably sour taste of this fungus, which you can try to remedy by marinating the slices before use.

Maki and nigiri sushi with Beefsteak Fungus

Believe it or not, one day you may manage to find so many good edible mushrooms that you simply can't eat them all. So, what to do?

Dried mushrooms

An easy way to store mushrooms is to dry them. This can be done by simply spreading a thin layer of mushrooms on a table on a sunny day, or by spreading them on a wire mesh on a radiator. The most effective, however, is to use a food dehydrator, where the mushrooms are put in narrow trays that can be stacked in many layers above a heat source.

In order to dry efficiently, it is important that the fruitbodies are not thick and fleshy. Thin-fleshed mushrooms, such as Horn of Plenty and Trumpet Chanterelles, can be placed directly in the dryer, while the more solid ones must be cut into slices to let the water evaporate.

When the dried mushrooms are put into jars and sealed properly, they will last for years. When the need arises, soak the dried fungi in water for an hour before cooking. Large boletes that tend to become slimy during frying may be much easier to prepare after drying.

An electric dehydrator and a basket full of Horn of Plenty ready for cleaning and drying

Trumpet Chanterelle (*Craterellus tubaeformis*) left to dry on a wire mesh

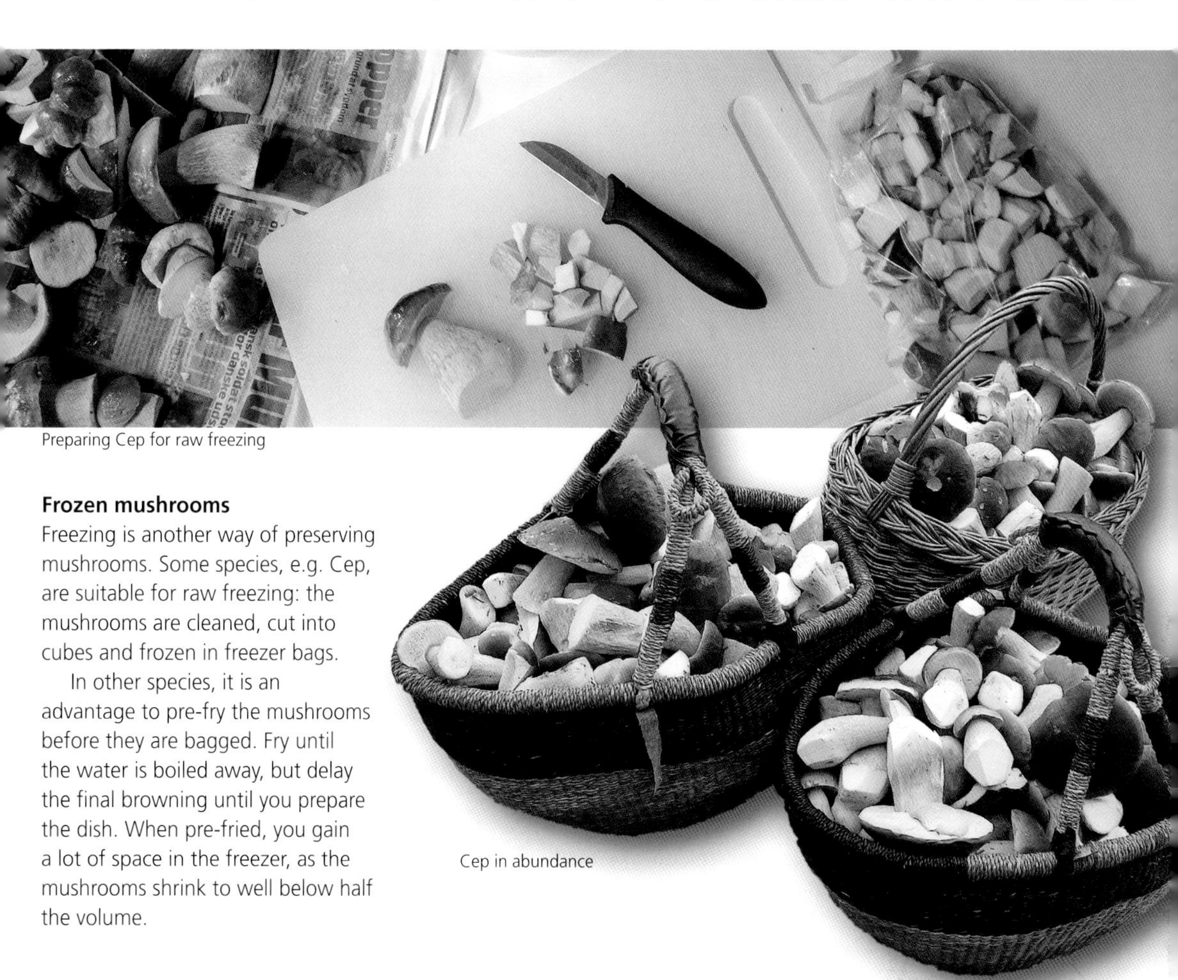

Preparing Cep for raw freezing

Frozen mushrooms

Freezing is another way of preserving mushrooms. Some species, e.g. Cep, are suitable for raw freezing: the mushrooms are cleaned, cut into cubes and frozen in freezer bags.

In other species, it is an advantage to pre-fry the mushrooms before they are bagged. Fry until the water is boiled away, but delay the final browning until you prepare the dish. When pre-fried, you gain a lot of space in the freezer, as the mushrooms shrink to well below half the volume.

Cep in abundance

Salted Woolly Milkcap (*Lactarius torminosus*)

Salted mushrooms

Salting mushrooms is a method of preservation that is used especially in Eastern Europe. An advantage of salting is that it can also remove the hot taste from bearded milkcaps (page 113). In Eastern Europe, these are therefore prized edible mushrooms.

Clean the milkcaps and soak them for a few minutes in boiling water. Drain and let them cool. Then put the mushrooms between layers of coarse salt in a washed and scalded jar. Press the pieces down with something heavy, such as a stone in a clean plastic bag. The salt will quickly extract the water so that the fungi are covered in brine. After a few weeks, the mushrooms are ready, but they need to be desalinated for a few hours in cold water before they can be used.

Mushroom-dyed yarn

Mushrooms are useful for much more than delicious cookery. They have a role in both hobby projects such as yarn-dyeing and industrially as biofactories in the production of medicines and enzymes. And as a tool in probably the oldest uses of all: alcohol production and as a raising agent for bread.

Dyeing yarn with fungi
The reason why fungi are coloured is a mystery, as the fungi don't need to attract insects for pollination or the like. Nevertheless, fungi can have very strong colours, including intense reds and blues. Many fungal dyes are unstable, and will disappear if you dry or heat the fruitbodies, but some species contain stable dyes that can bind to yarn. The Surprise Webcap (*Cortinarius semisanguineus*) for example will colour the yarn a splendid orange-red while the polypore fungus Cinnamon Bracket (*Hapalopilus nidulans*) will result in rare purple colours. You can make a fantastic palette from mushroom-dyed yarn!

C_2H_5OH
is the chemical formula for alcohol (ethanol). Alcohol is formed by fermentation, and this is where fungi, or more precisely yeasts, are essential.

When yeasts grow in an oxygen-free environment, they convert sugars into alcohol and carbon dioxide. The carbon dioxide bubbles away while the alcohol remains in the liquid, which may in this way be transformed into wine, beer or other beverages, depending on what sugars were used. If you close the bottle before the fermentation has stopped, the carbon dioxide gives a sparkling drink (e.g. beer or champagne).

Exactly the same process is used when raising bread. Here it is the carbon dioxide that is the essential part, as it forms the bubbles in the bread: the alcohol is a waste product.

Surprise Webcap produces orange-red colours

Biofactories

In modern biotechnological production, fungi play a major role as biofactories. Yeast fungi are easy to grow in an oxygen-free environment, and if you engineer the genetic code for a desired protein or enzyme into a yeast fungus, the fungus will excrete the desired substance during its growth. This can then be extracted from the fermentation tank.

Enzymes

In order to release nutrients, fungi have developed a large spectrum of enzymes. If you are looking for an enzyme that can enable washing powder to wash clean at lower temperatures or that can convert straw into biofuel, you should search among the fungal decomposers. When you find a species with a suitable enzyme, you can engineer the genetic code for that enzyme into a yeast fungus, which will then do the business.

Medicine

Finally, there are fungi with healing properties. Wild edible fungi, such as Hen of the Woods, have long been used in Asian medicine, typically to strengthen the immune system. But since World War II, fungi have also played a huge role in the West. Moulds, for example of the genus *Penicillium*, are in the main responsible for producing antibiotics. Due to these, people are much less likely to die from common infections such as pneumonia or blood poisoning.

As with enzyme production, the genetic code for an antibiotic is genetically engineered into a yeast fungus, which then produces the desired substance in a fermentation tank. The fungi are thus both source and biofactories.

Asexual reproduction in *Penicillium* – the genus that gave rise to the first antibiotic

The purpose of fungal fruitbodies is to undertake the sexual reproduction and to produce sexual spores that help the fungus to disperse.

Although fungal fruitbodies can appear very fleshy and well-structured, they are formed by tubular fungal hyphae which are only woven together, much like interwoven yarn.

The small, very young Ferny Bonnet (*Mycena pterigena*) on the right is magnified about 100 times. At this magnification you can clearly see the hyphae that make up the fruitbody.

Very young Ferny Bonnet on a stalk attached to a fern leaf – enlarged about 100 times so that the hyphae of the fruitbody are visible

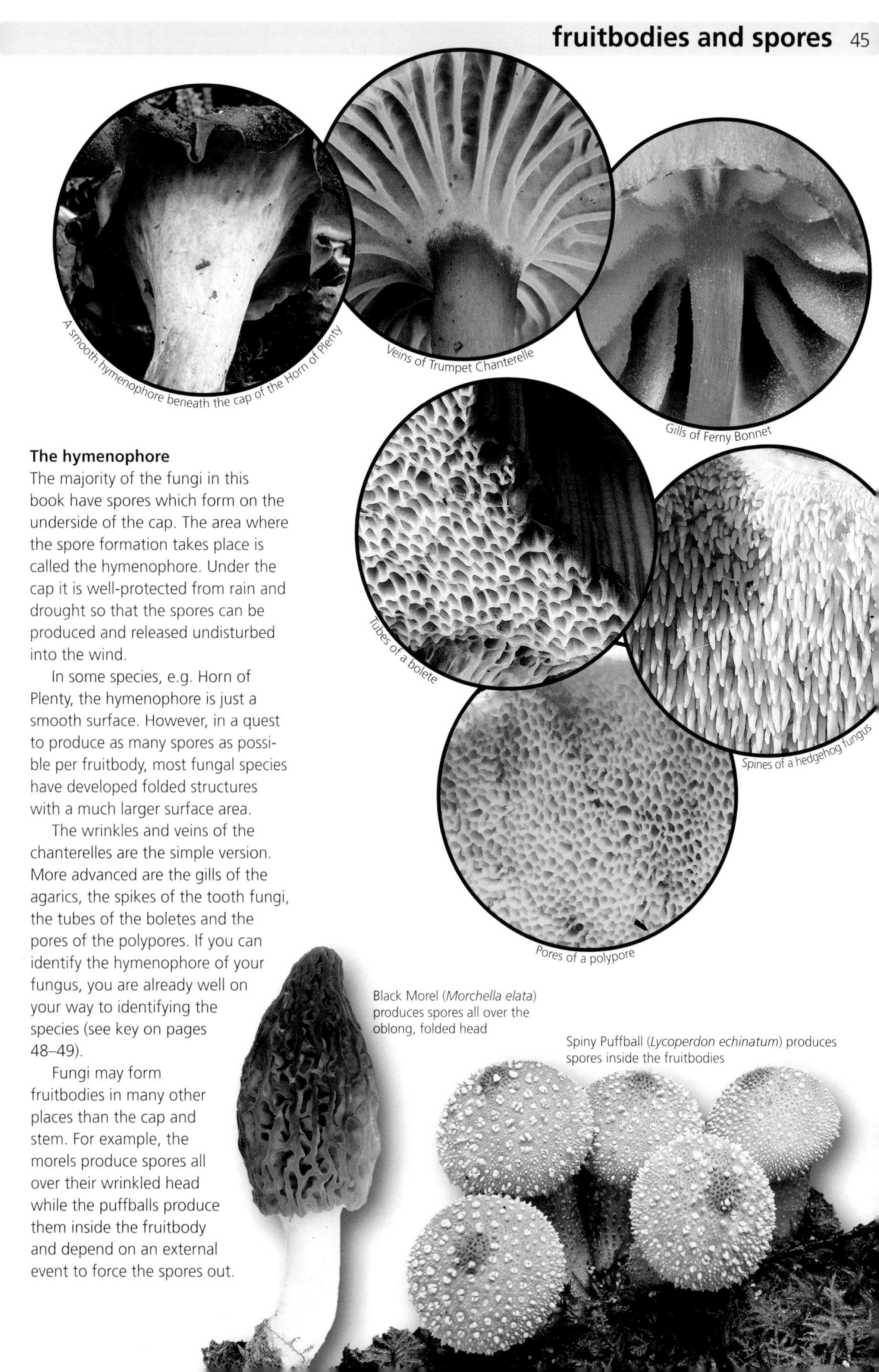

A smooth hymenophore beneath the cap of the Horn of Plenty

Veins of Trumpet Chanterelle

Gills of Ferny Bonnet

Tubes of a bolete

Spines of a hedgehog fungus

Pores of a polypore

Black Morel (*Morchella elata*) produces spores all over the oblong, folded head

Spiny Puffball (*Lycoperdon echinatum*) produces spores inside the fruitbodies

The hymenophore

The majority of the fungi in this book have spores which form on the underside of the cap. The area where the spore formation takes place is called the hymenophore. Under the cap it is well-protected from rain and drought so that the spores can be produced and released undisturbed into the wind.

In some species, e.g. Horn of Plenty, the hymenophore is just a smooth surface. However, in a quest to produce as many spores as possible per fruitbody, most fungal species have developed folded structures with a much larger surface area.

The wrinkles and veins of the chanterelles are the simple version. More advanced are the gills of the agarics, the spikes of the tooth fungi, the tubes of the boletes and the pores of the polypores. If you can identify the hymenophore of your fungus, you are already well on your way to identifying the species (see key on pages 48–49).

Fungi may form fruitbodies in many other places than the cap and stem. For example, the morels produce spores all over their wrinkled head while the puffballs produce them inside the fruitbody and depend on an external event to force the spores out.

Spore formation

Except for the morels, all of this book's edible mushrooms belong to the group of fungi called the Basidiomycota. In these, the spores are usually formed in fours at the top of a cell called a basidium. The basidia sit on the hymenophore, for example, on gills, on spines or inside tubes and pores. If you look at the picture below showing a gill at high magnification (20x–40x), you can see the spores sitting in groups of four at the top of the basidia.

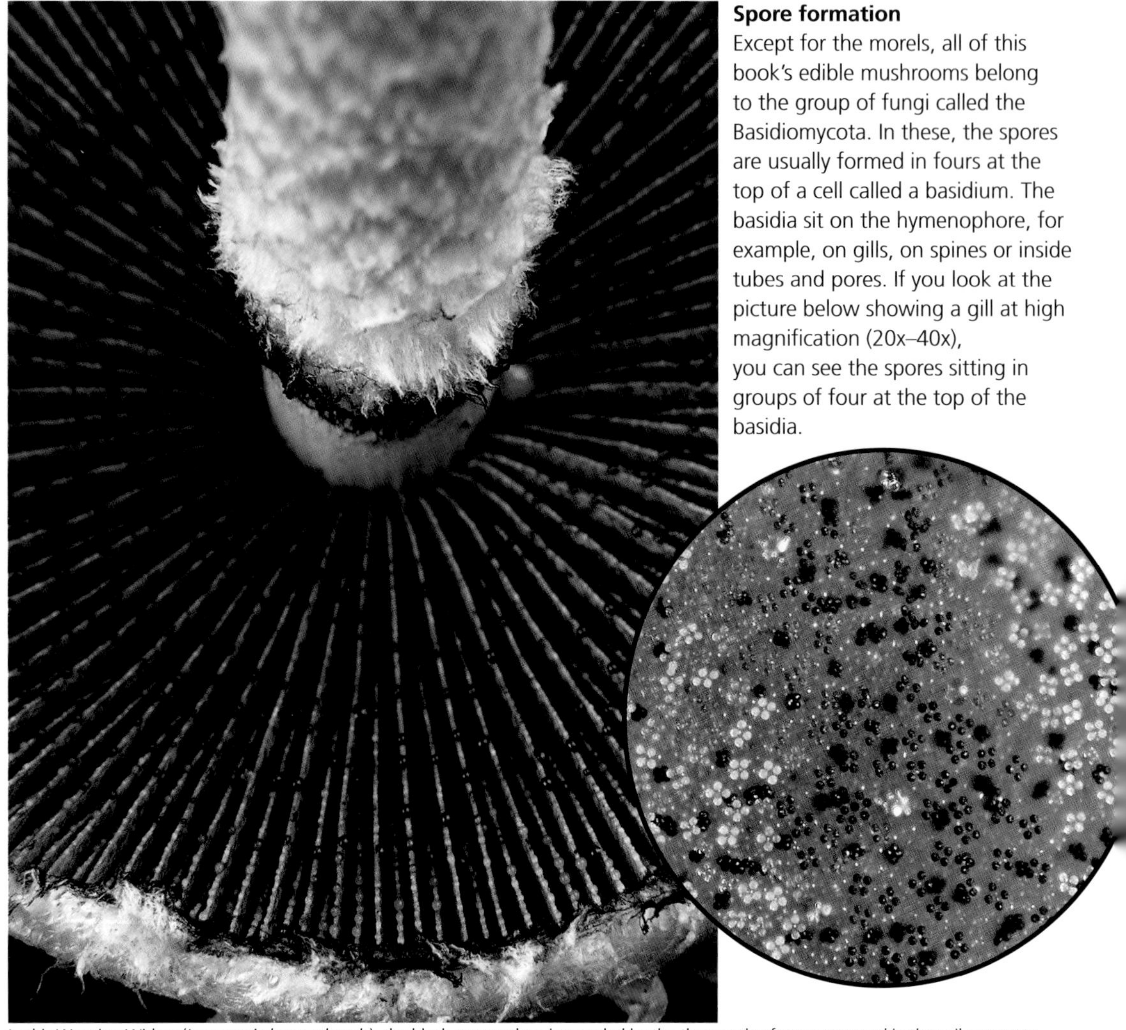

In this Weeping Widow (*Lacrymaria lacrymabunda*), the black spore colour is revealed by the thousands of spores trapped in the veil remnants – a close study of the gills (right) shows white (immature) and black (mature) spores sitting in fours at the top of the basidia

Spore colours

To identify an agaric, it is important to know the colour of its spores. However, spores are too small to see with the naked eye, so to be able to judge the colour, you have to look at a spore deposit.

If you have a good lens, you can examine the gills or search at the top of the stem where you can often find small spore deposits in wrinkles and remnants of the veil. If the fungi have grown in an overlapping fashion, there may also be spore deposits on top of the lower caps. But if nothing is found, you can make your own spore deposit.

The agarics are divided into these spore colours:

spore deposit whitish, cream to greyish, page 114

spore deposit brownish-pink to pink, page 132

spore deposit dark brown to black, page 134

spore deposit brown, page 144

How to make a spore deposit
Take a fresh cap and cut off the stem. Then place the cap with the gills facing downwards on a piece of white paper and cover it with a bowl or a piece of plastic to prevent it from drying. If the fungus is fresh and alive, it will usually release spores onto the paper within 2–4 hours.

START

ball- or pear-shaped, with dark spores inside when mature

puffballs, page 146

branched with acute tips

Upright Coral and Coral Tooth, page 55

folded

rather crisp

Wood Cauliflower, page 54

jelly-like to cartilaginous

jelly fungi, page 150

folded with a stem

morels, page 50

funnel-shaped, black

Horn of Plenty, page 94

spore deposit white, cream to greyish; gills white or coloured, page 114

spore deposit brownish-pink to pink; gills whitish to violet, page 132

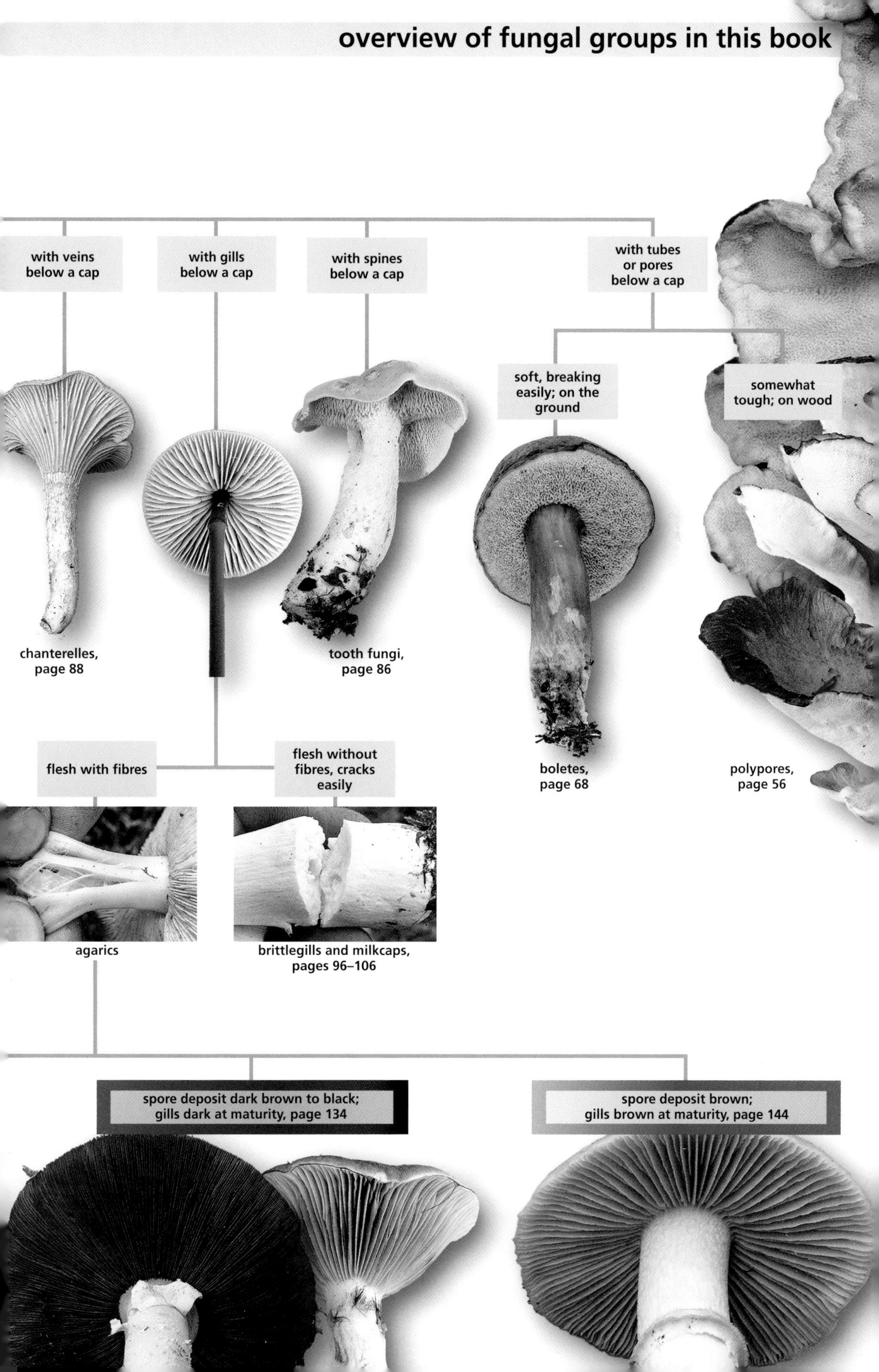
with veins
below a cap
with gills
below a cap
with spines
below a cap
with tubes
or pores
below a cap
soft, breaking
easily; on the
ground
somewhat
tough; on wood
chanterelles,
page 88
tooth fungi,
page 86
boletes,
page 68
polypores,
page 56
flesh with fibres
flesh without
fibres, cracks
easily
agarics
brittlegills and milkcaps,
pages 96–106
spore deposit dark brown to black;
gills dark at maturity, page 134
spore deposit brown;
gills brown at maturity, page 144

***Morels** are spring fungi that form large, almost club-shaped fruitbodies with a 'head' that has a honeycomb-like exterior. The whole fruitbody is hollow.*

OTHER SIMILAR FUNGI:
– in †† **False Morel**, the top is not hollow, but folded all the way through, p. 53.
– the **saddles** are likewise not entirely hollow but folded, p. 53.

Morels are among the most famous edible mushrooms. For example, morels were on the menu at the Danish Crown Prince Frederik and Crown Princess Mary's wedding in 2004.

A basket full of newly harvested Common Morels

•• **Common Morel** (*Morchella esculenta*) is the culinary best of the morels. It is a very distinctive edible mushroom with a chacteristic, powerful taste (enhanced by a few drops of lemon). Not to be eaten raw.

The Common Morel is immensely variable in shape and colour and can also grow in very different habitats (it is actuallly a complex of several closely related species which look more-or-less alike).

Typically forming fruitbodies during April and May, Common Morel is found mostly on soils with a high pH. Naturally alkaline soils on chalk or limestone can yield a stable harvest year after year. It can also be found around deporits of mortar and cement, although occurrences in such spots are likely to be sporadic. Common Morel lives in a mycorrhiza-like symbiosis with trees, perhaps especially elms.

Common Morel is hollow inside

•• **Semifree Morel** (*Morchella semilibera*) is a small, thin-fleshed and very slender morel. The margin around the lower part of the head is free of the stem, hence the name. It otherwise looks like a small Common Morel and is likewise a good edible mushroom, but you have to find quite a lot of fruitbodies for one dish. **It must not be eaten raw**.

Semifree Morel is found mostly in places affected by human activity, for example, where disturbance and earthworks have brought limy deposits to the surface. It fruits in April and May, at the same time as other species of Morel.

Semifree Morel in handfuls

•• **Black Morel** (*Morchella elata*) is less fleshy than the Common Morel. It has an elongated, pointed top, where the pattern is dominated by longitudinal ribs.

Unlike the Common Morel, Black Morel is a decomposer. One of the places where you can sometimes find it in large amounts is in beds of wood chips under, for example, roses or rhododendrons in parks and public gardens. In such places you can't count on a stable supply of morels year after year, but when you do find them, there may be hundreds of fruitbodies. Otherwise, Black Morel is often found in places where the soil has been disturbed, for example, in flower beds. Found April to May.

Black Morel is as tasty as the Common Morel. **It must not be eaten raw**.

†† **False Morel** (*Gyromitra esculenta*) is only distantly related to the true morels but looks a lot like them. The main difference is that the head of the False Morel does not have a large, internal cavity, but is instead folded in a brain-like manner.

Although deadly poisonous when raw, most of the toxins in False Morel are removed by drying or cooking. However, there are still some residues left, and it is therefore not advisable to experiment with eating species of *Gyromitra*.

False Morel forms a mycorrhizal association with pine on poor, gravelly or sandy soil. The fruitbodies are found from late March to May.

False Morel in a pine forest

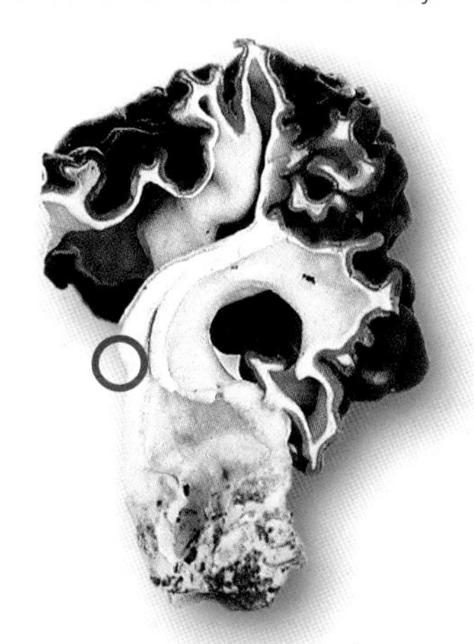

A sectioned False Morel is not entirely hollow

White Saddle by a roadside

• **White Saddle** (*Helvella crispa*) is another distant relative of the morels. It has a pitted or furrowed stem bearing a folded or saddle-shaped head. All parts are whitish.

White Saddle grows in humid hardwood forest soils, especially along the sides of roads and tracks, where it fruits during autumn. It is not poisonous, but not worth eating.

There are many other species of Saddle, such as the Elfin Saddle (*Helvella lacunosa*), which looks like a completely grey version of the White Saddle.

•• **Wood Cauliflower** (*Sparassis crispa*) grows on conifers. The fruitbodies can be as large as a football and consist of curly, flat, beige-coloured lobes. The flesh often has a slight chemical odour that disappears during cooking.

OTHER SIMILAR FUNGI:

– **Upright Coral** can have the same colour as the Wood Cauliflower but grows on hardwood. Its upward-pointing branches are cylindrical, not flat or lobed, see opposite.

– **Coral Tooth** forms large, branched, white fruitbodies lined with white, downward-pointing spines. It grows on old beeches and is edible but rare, see opposite.

Since the fruitbodies can be very large and keep fresh for a long time in the fridge, one fruitbody can be enough for many meals. The taste is fine and the texture crisp, but avoid old, brownish specimens as these may have a bitter taste. Try making a stew with cream and a little lemon, and mix it with pasta before serving.

Wood Cauliflower is found scattered in conifer plantations with spruce or pine. It usually forms fruitbodies at the base of the trees, often in the gap between two roots and can be found throughout autumn.

• **Upright Coral** (*Ramaria stricta*) forms medium-sized, orange to light brown, branched fruitbodies that grow from twigs and branches of beech. Unlike Wood Cauliflower, the branches are largely round in cross-section and the branching is upright.

Although it can be one of the most common fungi in an autumn beech wood, Upright Coral is of no value as an edible fungus.

There are many other species of coral fungi, some of which are edible. However, they look very much alike and should be avoided.

• **Coral Tooth** (*Hericium coralloides*) forms coral-like branched fruitbodies in which the white branches are clad with long, downward-pointing spines. It grows on old beeches and is mostly seen quite late in autumn.

Coral Tooth can be eaten, but since it is rare and registered as Near Threatened on the British Red Data list, you should leave it to beautify the woodland scene.

Polypores *form fruitbodies where the underside is covered in small holes (pores). The pore lining produces the spores, which disperse in the air when they drop out of the mouth of the pore. They mostly grow on wood.*

OTHER SIMILAR FUNGI:

– **boletes** also have holes below the cap, but these consist of single tubes that can usually be separated from each other and from the flesh of the cap. Almost all boletes grow on the ground.

The fruitbodies of the polypores range in size from small to huge. Some species may form perennial fruitbodies as hard as wood and the largest are measured in metres and multiple kilos. The edible polypore species included in this book are, however, all quite soft-fleshed and have a stem or narrowed base, whereas more typical polypores, such as the Tinder Bracket (*Fomes fomentarius*), are broadly attached to the wood.

Tinder Bracket forms hard, broadly attached fruitbodies

Dryad's Saddle seen from below. The large pores are clearly visible

Caps with coarse scales

• **Dryad's Saddle** (*Cerioporus squamosus*) forms very large, rather soft-fleshed fruitbodies, which are attached to the tree with a short, dark stem. The upper side has coarse scales and the pores are several millimetres wide.

OTHER SIMILAR FUNGI:
– **Tuberous Polypore** (*Polyporus tuberaster*) is similar but smaller. It is also edible.

Dryad's Saddle is common on old deciduous trees, especially in parks and gardens. It forms fruitbodies mostly during May and June, but may also be found later in the year. Young specimens are edible and smell a bit like cucumber or watermelon. The stem and the mature fruitbodies are tough and cannot be used.

Dryad's Saddle is not one of our top edible fungi. It has, however, recently entered the menus of gourmet restaurants as part of the wave of New Nordic Cuisine.

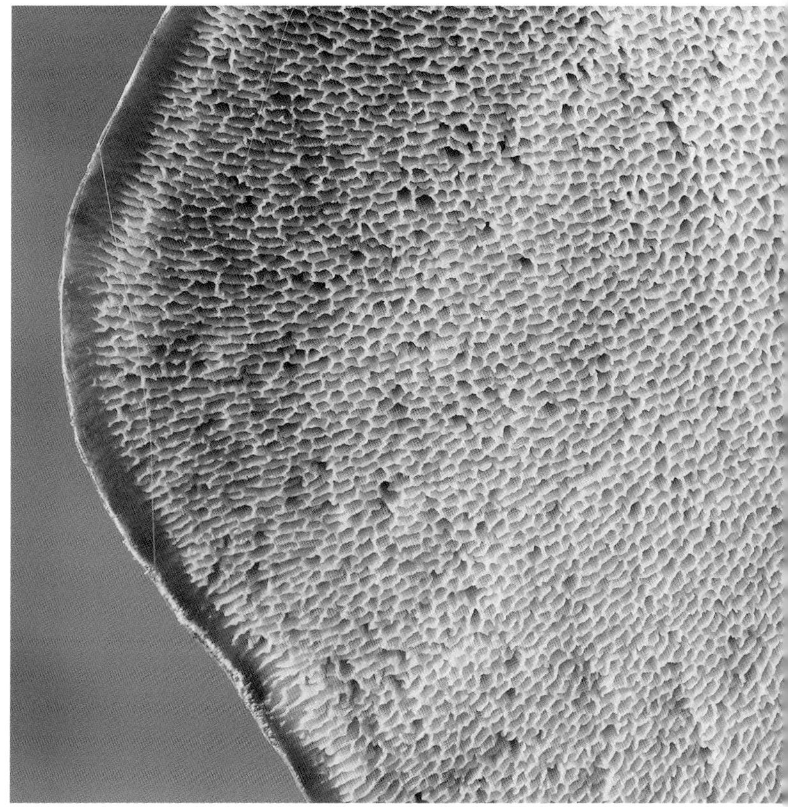
Very large, angular and somewhat oblong pores

• **Chicken of the Woods** (*Laetiporus sulphureus*) is a very large, orange-yellow polypore that form magnificent tiered clusters on living trees. Some individuals develop fruitbodies as early as late May, while others appear during autumn. It is most common on oak, but can also be found on a large number of other deciduous trees.

OTHER SIMILAR FUNGI:

– the even larger, also edible **Giant Polypore** (*Meripilus giganteus*) has the same basic shape. It is brown with a beige pore surface, which turns dark when bruised (see page 62). It grows on old beeches.

Young Chicken of the Woods, where the pores are not yet properly developed, are edible after thorough cooking. Cut them into thin slices or small cubes and fry them in a pan (they take up a lot of oil). Make sure the pieces are heated all the way through, as raw or poorly cooked Chicken of the Woods may cause an upset stomach.

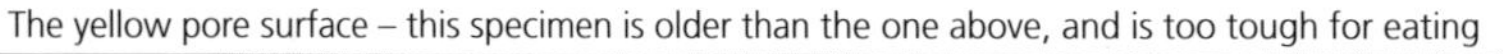

The yellow pore surface – this specimen is older than the one above, and is too tough for eating

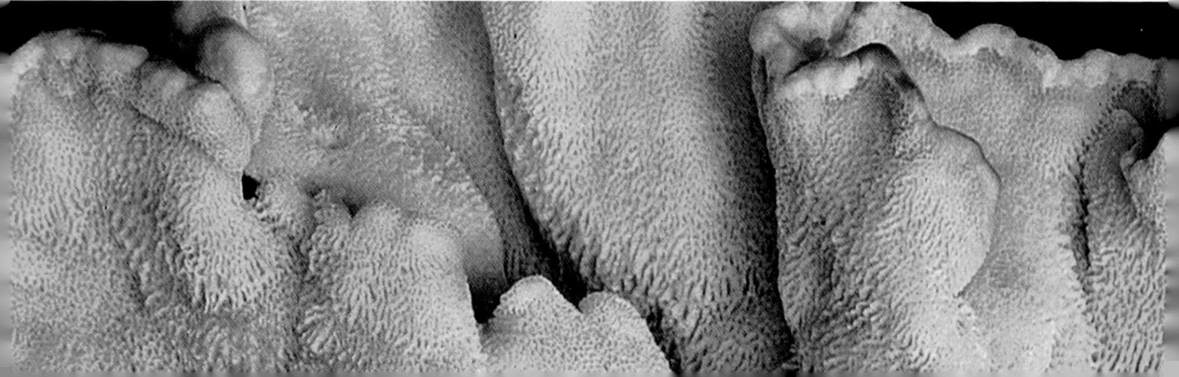

Chicken of the Woods tastes mild and has a very firm consistency that is quite reminiscent of chicken meat. Some say that specimens that have grown on oak taste more sour than those that have grown on other deciduous trees.

Various species of Chicken of the Woods are very popular edible fungi around the world. In Bhutan, where the species goes by the local name 'Taa Shamong', it is offered from large plastic bags at the Thimphu market.

Stewed Chicken of the Woods with green pasta and basil

•• **Hen of the Woods** (*Grifola frondosa*) forms large, tiered fruitbodies with many small caps. Each brown-grey cap looks like a small tongue, attached by a stalk to the rest of the fruitbody. The pore surface is white and the pores are quite large. Hen of the Woods grows on old oaks, where it typically emerges in a gap between two roots during autumn (see also page 17).

OTHER SIMILAR FUNGI:

– **Giant Polypore** (*Meripilus giganteus*) has similar colours, but its pore layer darkens when touched; also, it is much larger and grows on beech (p. 62).

– **Umbrella Polypore** is very similar, but here the small caps are attached with a central stem (see opposite).

Hen of the Woods is an excellent edible fungus with a fine consistency and better taste than the other edible polypores. It is quite common in the southern parts of the UK. The species can also be grown on sawdust. In China and Japan, it is considered an important ingredient in traditional medicine, and studies show that it stimulates the immune system. Its Japanese name is 'Maitake'.

The undersides of the small caps are covered in white pores

A mature Hen of the Woods gnawed by snails and beetles but still an excellent edible fungus

• **Umbrella Polypore** (*Cladomeris umbellatus*) resembles Hen of the Woods, but its small caps is attached by a central stem. It grows on the ground near old beeches, but is also occasionally found by oaks.

Umbrella Polypore is edible, but less tasty and much softer than Hen of the Woods, and is also often attacked by insect larvae. It fruits rather early, from July to September.

• **Giant Polypore** (*Meripilus giganteus*) is one of our most striking fungi, forming gigantic, tiered, brown fruitbodies at the base of beech trees. It lacks the bright yellow and orange colours of Chicken of the Woods and its beige-coloured pore surface becomes darker when touched.

Giant Polypore likes to grow on old beeches in urban and suburban landscapes, for example, in parks, avenues and cemeteries. It decomposes the wood and eventually the tree dies. The fruitbodies typically appear in August and early September.

OTHER SIMILAR FUNGI:

– **Chicken of the Woods** (*Laetiporus sulphureus*) has almost the same size and shape, but is bright yellow to orange, see page 58.

– **Hen of the Woods** (*Grifola frondosa*), which is also edible, has similar coloration but is smaller. It does not turn dark when touched and grows on oak, see page 60.

A giant, inedible fruitbody with many layers of caps

Young and edible fruitbodies

There have been heated discussions about the edibility of the Giant Polypore. It is not poisonous, but its somewhat sour, slightly perfumed taste is not what most mushroom hunters expect from a good edible fungus. This, however, did not stop the mycology students in the photograph from helping themselves! Personally, I think young fruitbodies are quite OK. And as a special little twist, the whole dish will darken during cooking, and ends up almost black – deliciously scary!

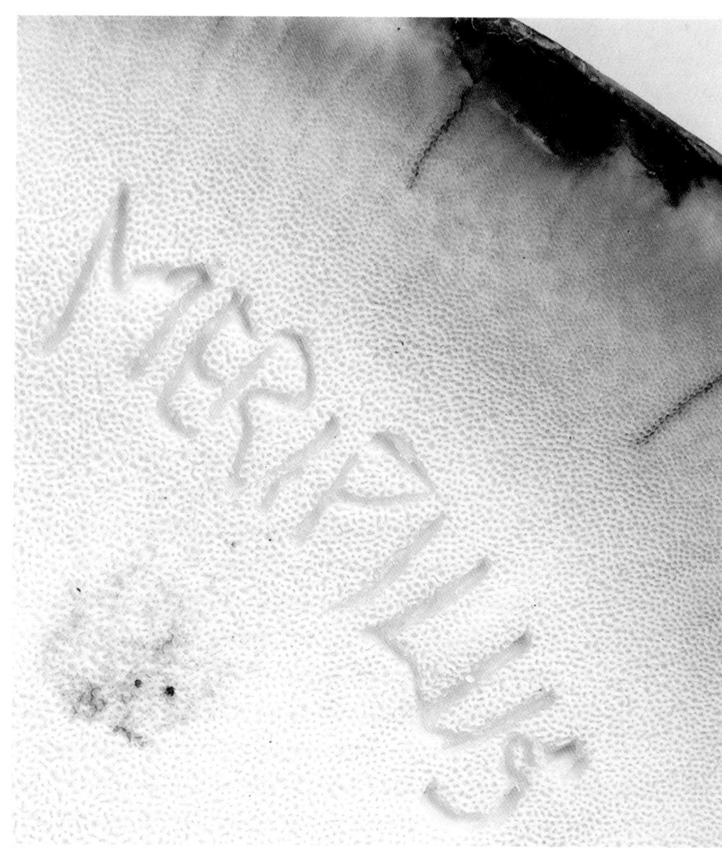

The pore surface becomes dark when touched

A rather old specimen, but fortunately mycology students are optimists

• **Beefsteak Fungus** (*Fistulina hepatica*) forms peculiar, tongue-shaped fruitbodies filled with a red juice. When cut, a juicy, structured flesh with the texture of raw tongue is exposed. The underside of the fungus is covered with small tubes. In a young specimen these are fused together and the mouths of the tubes are closed, but as it matures the pores transform into free, open tubes. The fruitbodies may have a short stem (see the picture at the bottom of page 68).

Growing only on old oaks, Beefsteak Fungus typically appears in August or September.

OTHER SIMILAR FUNGI:
There are no other polypores with red juice.

Young Beefsteak Fungus

Beefsteak Fungus is edible, but has a very special, sour taste, and there are divided opinions about its gastronomic value. It is one of the few species that may be used raw. You can try it in salads or pickles, but the most spectacular use is probably on Nigiri sushi. For this, always use young specimens and look forward to a unique culinary experience!

Nigiri sushi with raw, marinated Beefsteak Fungus, roasted sesame seeds and wasabi

An older fruitbody, showing red juice and free tubes when cut

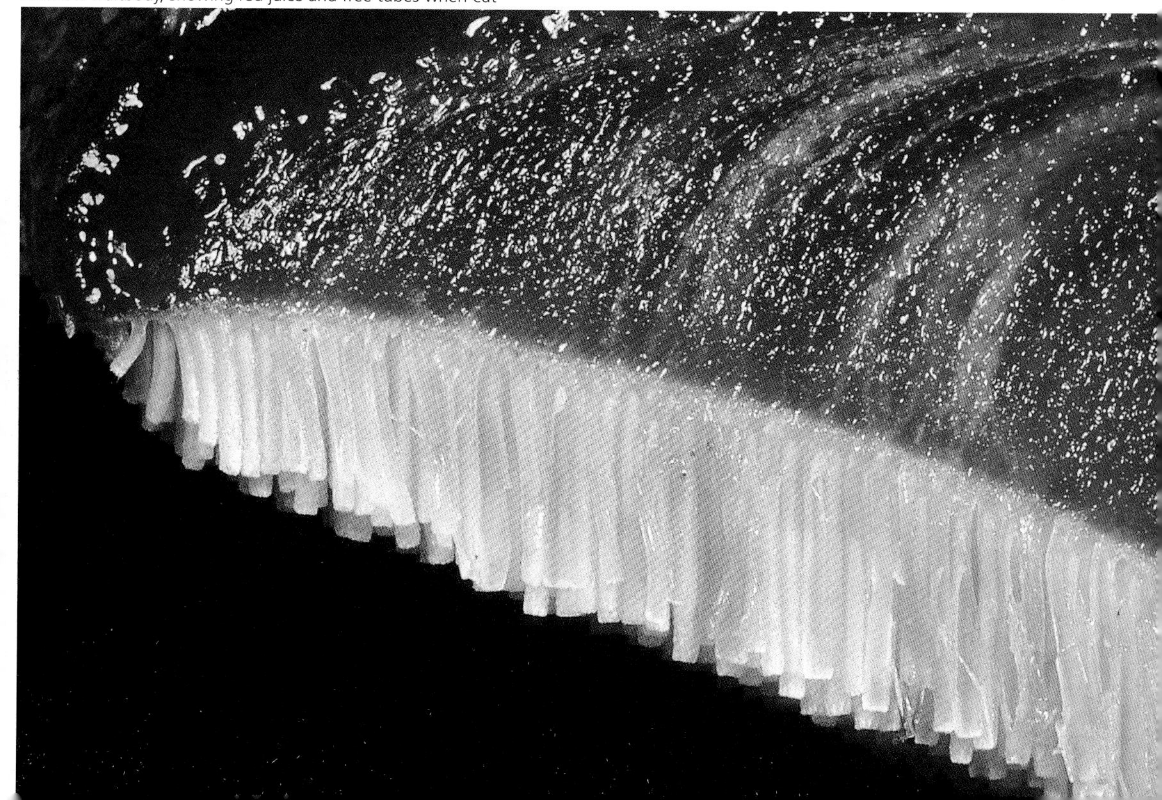

•• **Forest Lamb** (*Albatrellus ovinus*) is a polypore that grows on the ground. Its fruitbodies resemble very pale boletes but have a pore layer that runs down the stem. The cap is very pale brown, while the pore mouths are white, changing colour to pale yellow when touched. When fried, all parts turn yellow.

OTHER SIMILAR FUNGI:

– **Fused Polypore** (not known from Britain and Ireland) is very similar, but it has a browner cap, the pore mouths do not turn yellow when touched or fried, and it tastes bitter, see opposite.

– **boletes** may have the same shape, but are usually darker and do not turn yellow when touched, see page 68.

– species of **Falsebolete** (*Boletopsis*) have a more greyish colour, taste bitter and do not turn yellow when touched. They are not poisonous, but they are very rare and should not be collected.

Forest Lamb turning yellowish when bruised (top image), and a darker shade when fried (below)

Forest Lamb fruits in autumn in old conifer forests, especially under spruce on calcareous soils.

As it is extremely rare in Britain and Ireland, you have a much better chance of meeting the lamb in the forests of Norway, Sweden and the Alps. Here you will discover that it is an excellent edible fungus. As usual, only young specimens should be collected and even these should be thoroughly cooked. Well-fried, you will get a meal with a fine taste and a delicate consistency.

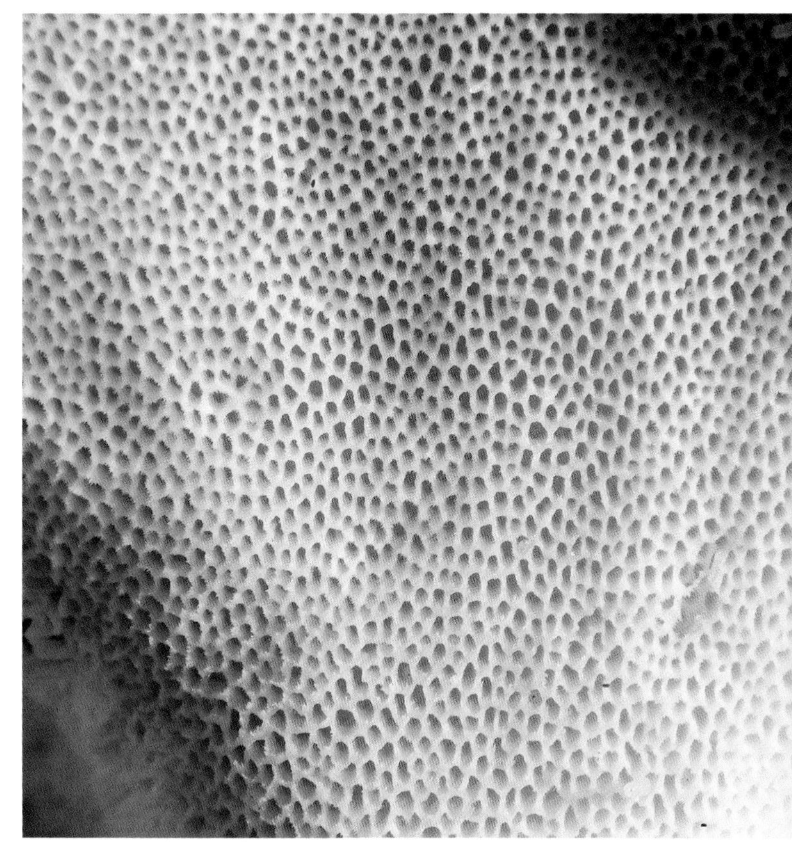

Forest Lamb has white, angular pores

• **Fused Polypore** (*Albatrellus confluens*) resembles Forest Lamb, but has a somewhat browner cap and does not turn yellow when touched. It tastes bitter and should not be mixed with edible fungi, although it is probably not poisonous. Despite being widespread in European conifer forests, Fused Polypore is not known to occur in Britain and Ireland.

***Boletes** form soft-fleshed fruitbodies with a cap and a stem. Beneath the cap is a layer of hundreds of small tubes within which the spores are formed. The tubes appear to form a solid tube layer, but they can usually be separated from each other into individual tubes and they can also be separated quite easily from the cap flesh.*

Since the boletes are very easy to recognize as a group, they are really good beginners' mushrooms. There are a few moderately poisonous species (page 81) and a few with a bitter taste (page 77), but the majority are edible and Cep, Scarletina Bolete and Lurid Bolete are among the very best edible mushrooms.

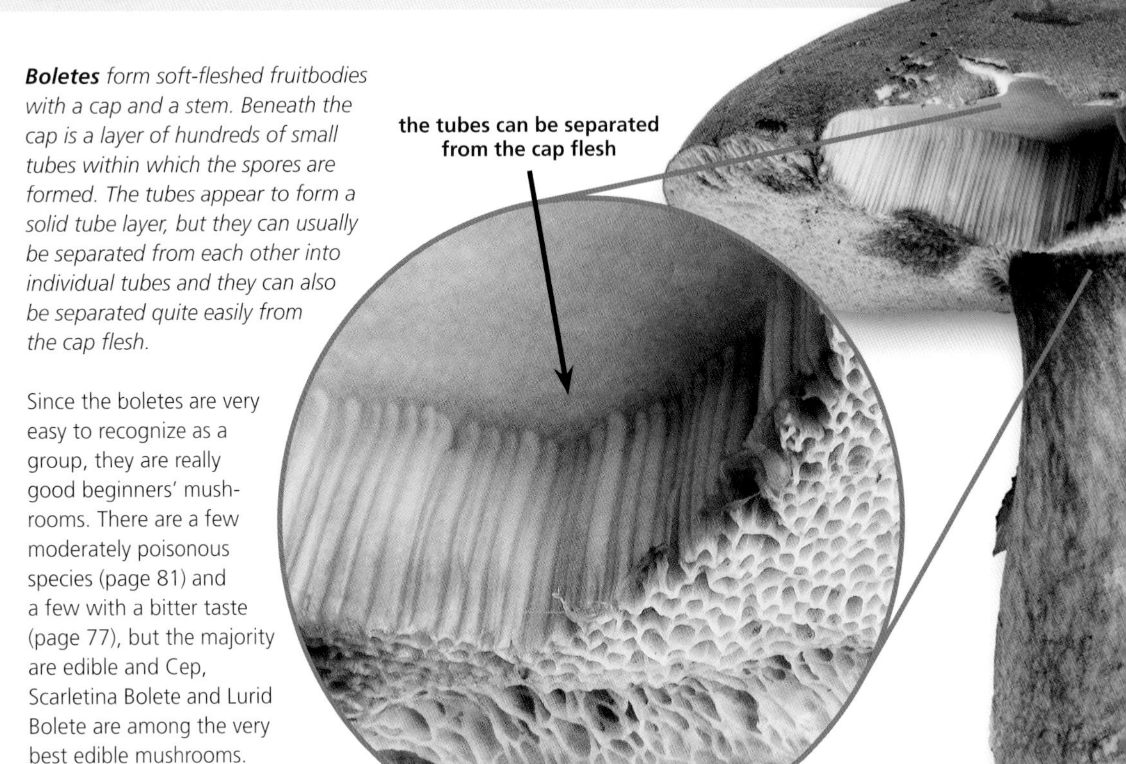

OTHER SIMILAR GROUPS OF FUNGI:

– **Dryad's Saddle** and similar polypores also have a cap and a stem. Their flesh is rather tough and their pores are inseparable from each other and from the flesh of the cap. They mostly grow on wood, see page 57.

– **Forest Lamb** is soft-fleshed with a bolete-like shape and grows on the ground. However, it has the typically inseparable pores of a polypore, see page 66. This and related species are not poisonous.

– **Beefsteak Fungus** is soft-fleshed and has separable tubes like boletes. It grows on wood and is filled with reddish juice, see page 64. It is edible.

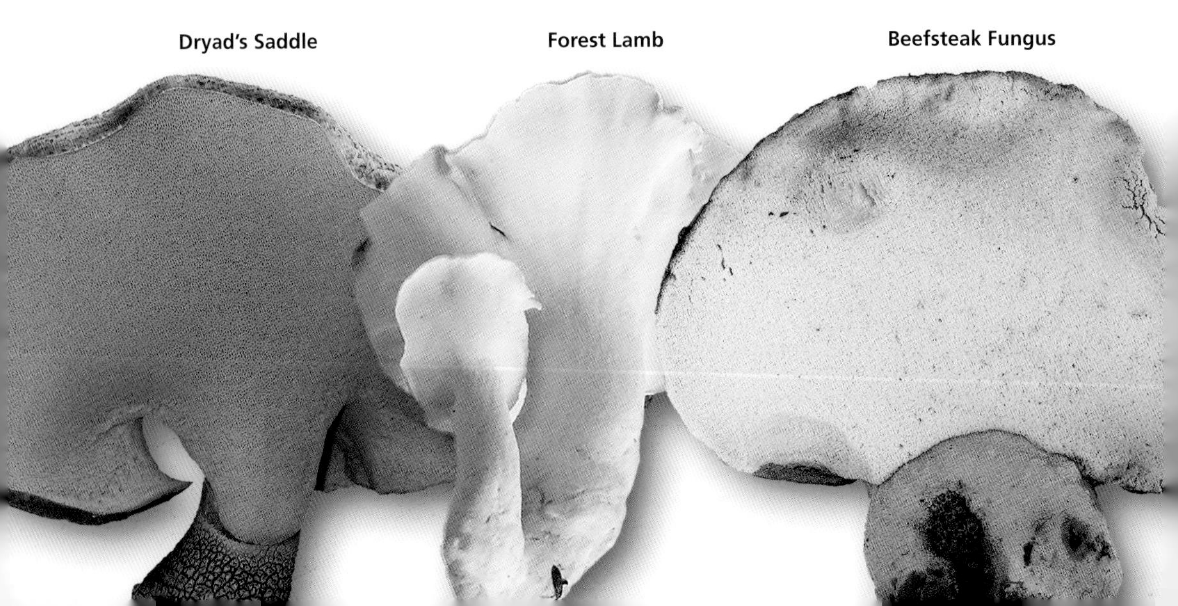

flesh
tubes
tube mouths
boletes
– see the various groups
on the next spread

Preparation of boletes
Boletes contain a lot of water, and since some species can also be a bit slimy, it is a bit of an art to fry boletes and retain some structure. Thus, one should not be tempted to pick a lot of old, soft boletes. Stick with the young, firm fruitbodies, even though it may be difficult to resist picking the older ones, too.

To cook, place a single layer of mushrooms in the pan at a time, cook over a high heat and let the liquid boil away. Then add a little oil or butter and fry until the mushrooms become crispy at the edges.

A variation is to slice young Ceps very thinly and fast-fry them in plenty of very hot oil. In this way, the fungi hardly shrink and they keep some of their nut-like raw taste. Unfortunately only a few slices can be processed at once, so a large meal will take time; fortunately, it will be worth it in the end!

Roughly cleaned Lurid Boletes (*Suillellus luridus*)

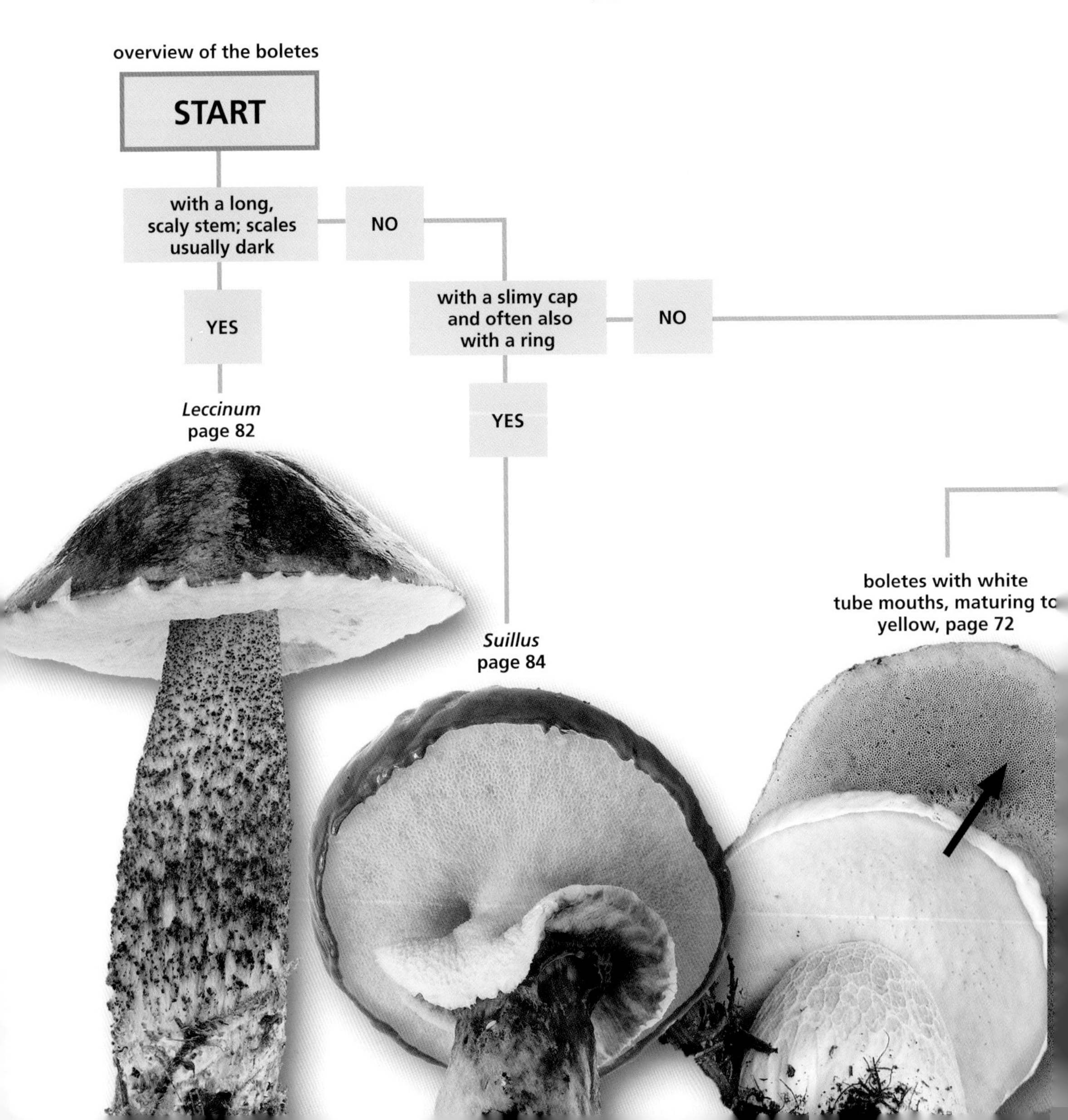

If you find a lot of boletes, you can dry or freeze them and save them for leaner times, see page 40.

Drying
Since boletes are very fleshy, it is important that they are cut into thin slices before drying. It takes time to dry boletes, but they become easier to use afterwards because they become less slimy during cooking.

Freezing
Boletes are very suitable for freezing. You can either freeze the raw mushrooms as small cubes or begin by cooking off the excess water first so they will take up less space.

Finished stew of Lurid Bolete

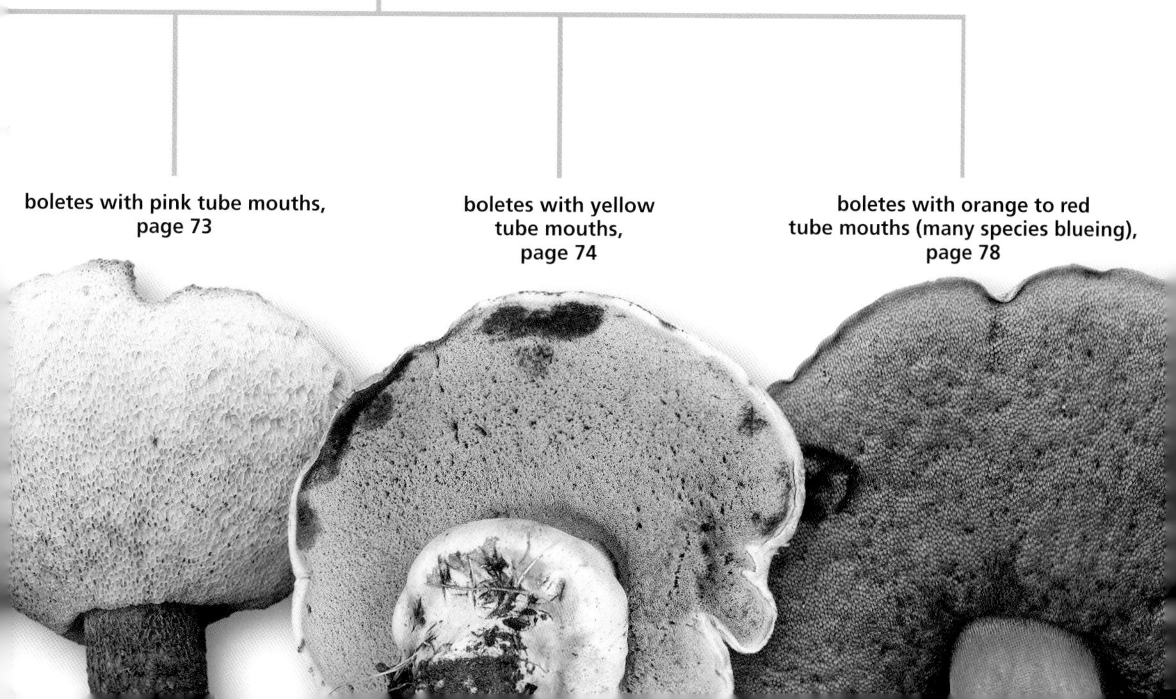

boletes with white tube mouths, maturing to yellow

The champagne cork stage of Cep

White tube mouths and pale net ornamentation

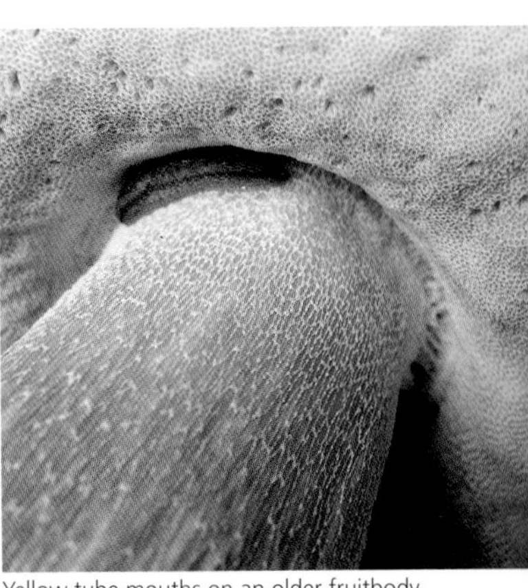
Yellow tube mouths on an older fruitbody

Boletes whose tubes start out completely white but turn yellow as they mature belong to the ***cep group****. They have pale net-like ornamentation down the stem and are all edible and good.*

OTHER SIMILAR FUNGI:

– fungi in the bolete genus ***Leccinum*** have long, scaly stems and tubes that remain whitish to greyish, see page 82.

– **Forest Lamb** (*Albatrellus ovinus*) has white tubes that cannot easily be detached from the cap, see page 66 (rarely found in Britain and Ireland, but commoner in, for example, Scandinavian coniferous forests).

Species of the cep group of boletes form large fruitbodies with a club-shaped stem and a rounded cap, shaped like a bun. They never turn blue when touched or cut.

You have to start looking for members of this group from the end of August or the beginning of September; timing varies according to how dry the summer has been. They then may fruit in a quite concentrated period of two to three weeks, the so-called cep boom. Ceps can still be found later in the autumn, but are then usually much more scattered.

Cep should be collected young. As long as they are shaped like champagne corks, they are among the best edible mushrooms, but as soon as the tubes start to turn yellow they become soft. If you want to use middle-aged fruitbodies with yellow tubes, remove the tubes before cooking or the dish will become very sticky. Old fruitbodies should be avoided.

Species of the cep group form mycorrhizal associations with many different trees, e.g. beech, oak, birch, spruce, pine and fir. If you want to distinguish between the species, you must know the host tree and study the details of the cap cuticle. Fortunately, all species are equally tasty.

•• The main species of the group is **Cep** (*Boletus edulis*), also called Penny Bun. It can be recognized by its greasy cap surface. If you try to tear the cap cuticle, you will find that it is very tough and rather jelly-like.

Cep grows in many types of woodland, such as under beech, spruce, pine and fir. It is the most common of the species in its group and can be found throughout autumn.

•• **Summer Bolete** (*Boletus reticulatus*) is another fairly common species. It has a paler cap than Cep and the cap cuticle has a dry, matt texture. It is found with beech and oak, and prefers open landscapes such as parks, on good soil. As its name suggests, it is most common from summer and into September.

• **Bitter Bolete** (*Tylopilus felleus*) is rather similar to Cep, but is inedible. It can be identified by the white tubes that turn pink with age, the dark net ornamentation of the stem and the very bitter taste. It is not poisonous, but a single fruitbody can taint an entire dish.

Bitter Bolete is quite common in August and September, and grows with both deciduous and coniferous trees.

*Small to medium-sized **boletes** with yellow tube mouths. Most species lack well-defined net ornamentation on the stem. This group has no poisonous members, but several are bitter-tasting and some are soft-fleshed, best suited as a supplement to fungi with a firmer texture.*

OTHER SIMILAR FUNGI:

– the two bitter-tasting and inedible species, **Rooting Bolete** (*Caloboletus radicans*) and **Bitter Beech Bolete** (*Caloboletus calopus*), have a fine-meshed net on the stem, see page 77.

– species of the genus *Suillus* all have a somewhat slimy cap surface and stem base and often also have a slimy ring around the stem. All are edible, but not very good, see page 84.

• **Bay Bolete** (*Imleria badia*) is a medium-sized bolete with a brown, smooth stem and yellow tubes. The cap is at first felted (like suede) but later becomes sticky, especially in warm, humid weather. The tube mouths are small and turn blue a few seconds after bruising. Bay Bolete can grow with both deciduous and coniferous trees, but is most common in spruce forests.

Bay Bolete is one of the better edible mushrooms in this group. Young specimens may have almost the consistency of Cep, but not as good a taste.

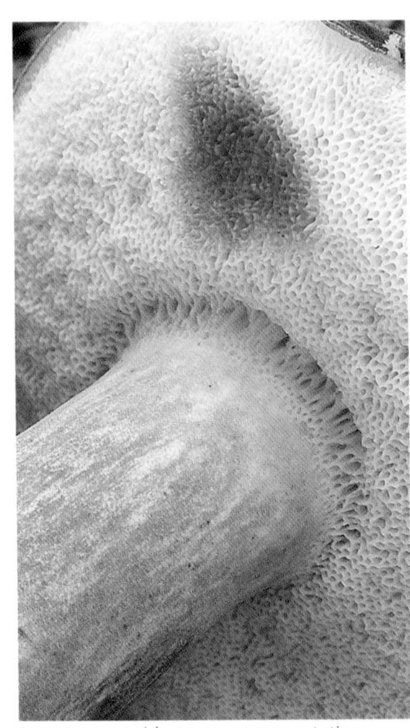

Brown stem without net ornamentation and tube mouths turning blue

• **Rusty Bolete** (*Xerocomus ferrugineus*) can be recognized by its felted cap (like suede) and the coarse, bright yellow tube mouths, which usually do not become blue when bruised. Often the tubes continue a little down the stem and here form a short, coarse net ornamentation. The stem base usually has bright yellow hyphal strings.

This species belongs to a group of very soft-fleshed boletes and should only be used in combination with more tasty and hard-fleshed fungi. It grows in both deciduous and coniferous forests. There are several very similar relatives with similar culinary characteristics, e.g. Suede Bolete (*Xerocomus subtomentosus*).

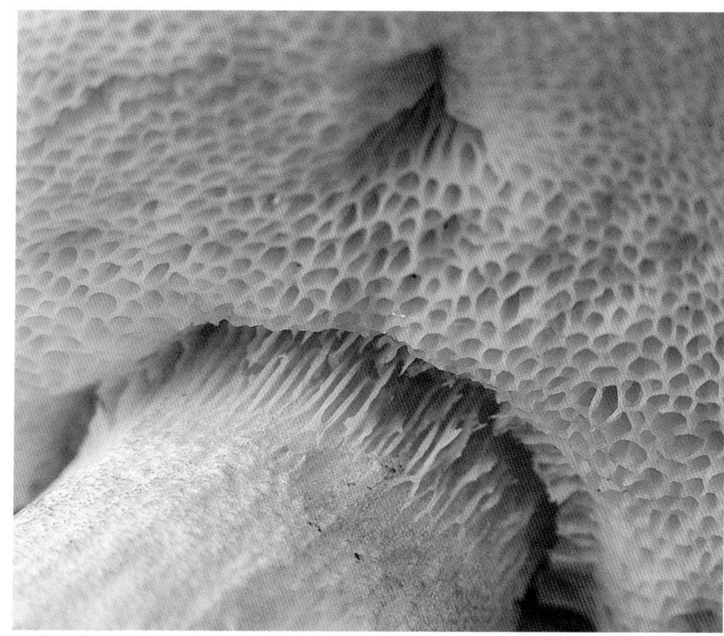

Bright yellow, coarse tube mouths and the suggestion of a net ornamentation

The genus **Xerocomellus** *contains a number of small boletes with stems 1–2 cm thick and yellow tube mouths, and without well-developed net ornamentation.*

The species of the group are difficult to separate. As they do not have much taste and are all rather soft-fleshed, they are only considered moderately good edible mushrooms – best suited as supplementary to other fungi.

• **Red Cracking Bolete** (*Xerocomellus chrysenteron*) forms rather small, soft-fleshed fruitbodies with a red-dotted stem, pale yellow tube mouths and a brown, cracking cap. It grows with beech and conifers and is mostly found from August to September.

Matt Bolete is very similar, but usually fruit during September and October, see below.

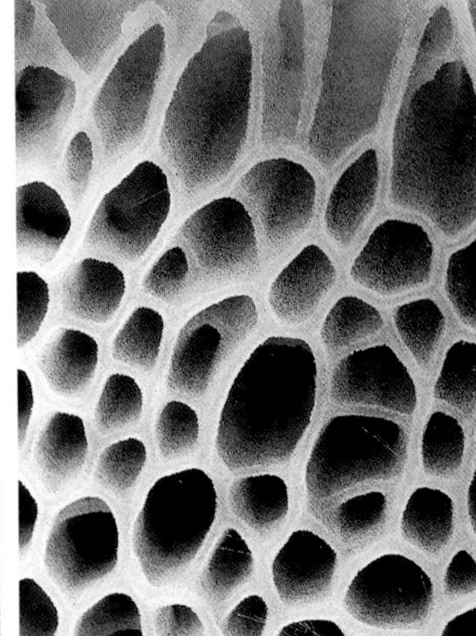

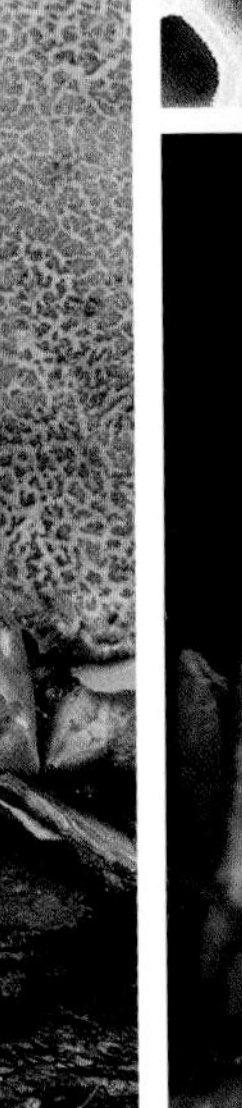

Red Cracking Bolete with cracking cap and red-dotted stem

• **Matt Bolete** (*Xerocomellus pruinatus*) is extremely similar to Red Cracking Bolete (above), but its cap shows fewer cracks and often has a narrow, red zone along the margin. Young specimens may have a characteristic dew-like cap surface.

Matt Bolete mostly grows in beech forests and often fruits later than Red Cracking Bolete, typically in late September and October. There are many similar species, all soft-fleshed and of secondary culinary value.

*There are two inedible bitter boletes with yellow tubes, namely **Bitter Beech Bolete** and **Rooting Bolete**. Both are large boletes with a net ornamentation on a stem that is more than 2 cm thick.*

• **Bitter Beech Bolete** (*Caloboletus calopus*) forms pale-capped fruitbodies with pale yellow tube mouths and a stem with unusually beautiful net ornamentation. At the top the net is yellow, but towards the base it changes gradually to red. All parts of the fruitbodies turn blue when wounded and the flesh tastes very bitter.

Bitter Beech Bolete is found scattered on nutrient-poor soil in both deciduous and coniferous woodlands.

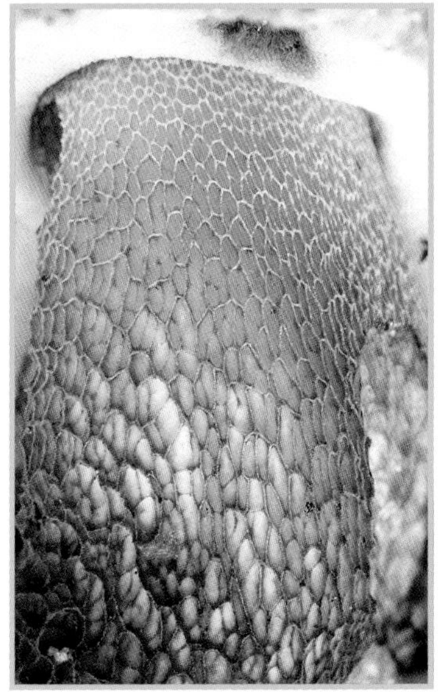

The beautiful stem of Bitter Beech Bolete

• **Rooting Bolete** (*Caloboletus radicans*) forms very large fruitbodies with pale caps and pale yellow colours on the tubes and stem. Its stem has a very fine, yellow net ornamentation that can be difficult to see without a lens. The flesh and tube mouths turn blue when bruised, and it has a bitter taste.

Rooting Bolete is found scattered with deciduous trees in warm parks and cemeteries. It fruits mostly in August and the first half of September.

There are several very similar but mild-tasting species in the genus *Butyriboletus*, e.g. the rare Pale Bolete (*Butyriboletus fechtneri*) and Oak Bolete (*Butyriboletus appendiculatus*). These are all edible.

A very fine-meshed net on Rooting Bolete

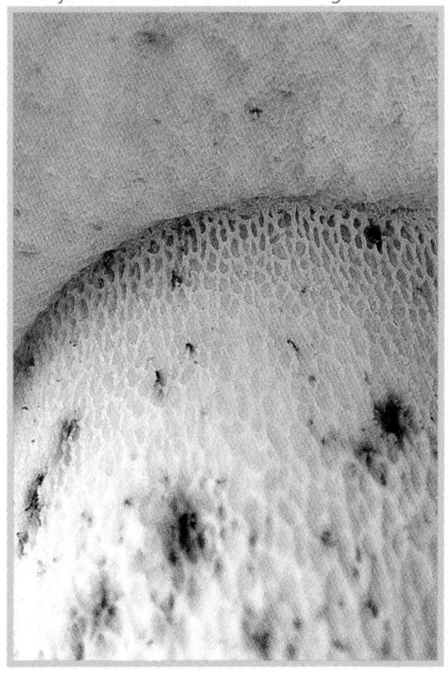

The genera **Suillellus** *and* **Neoboletus** *contain moderately large boletes with red tube mouths. Stem, flesh and tubes will within seconds change colour from yellow to indigo blue when cut or bruised. All species are edible, but only after cooking;* ***eaten raw, they may cause poisoning, resulting in dizziness and a stomach upset****.*

OTHER SIMILAR FUNGI:

– † **Devil's Bolete** has a light grey cap and reacts moderately light blue when cut, see page 81.

– † **Bilious Bolete** also has a pale cap, but with distinct pink hues. It also turns moderately light blue when cut, see page 81.

– **Peppery Bolete** is smaller with yellow flesh that does not turn blue, see page 85.

Scarletina Bolete with yellow but blueing tubes and red tube mouths

•• **Scarletina Bolete** (*Neoboletus erythropus*) forms large, firm fruitbodies and is an excellent edible fungus (but only when cooked). It has a brown or yellow-brown cap and is differentiated from similar boletes by the stem surface which does not have net ornamentation, but is instead covered with small, orange-red dots.

The species is common in deciduous and coniferous woods on slightly poor soils. It is most frequent in August and September.

A close-up of the dotted stem surface

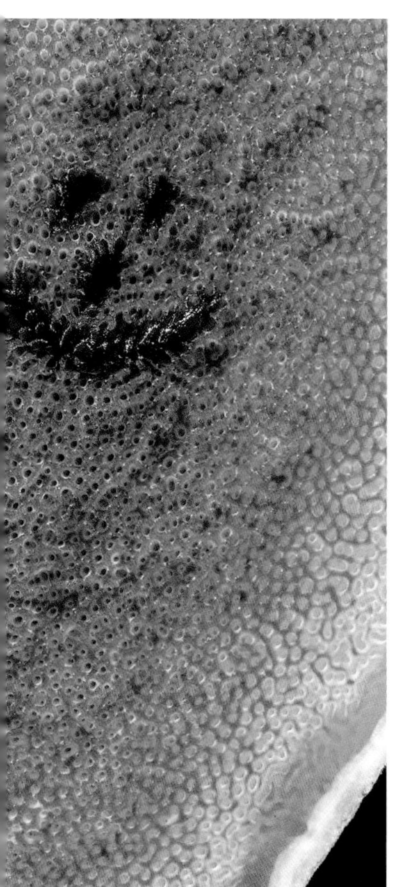

Reddish, strongly blueing tube mouths

Young fruitbodies of Scarletina Bolete

•• **Lurid Bolete** (*Suillellus luridus*) forms large fruitbodies with a light brown cap and a thick stem that has an orange net ornamentation. All parts turn blue very quickly and vigorously. Uniquely, the surface where the yellow tubes are attached to the cap flesh is red (see the picture to the right).

The species is one of our very best edible mushrooms – **just never eat it raw, as it may cause dizziness and an upset stomach**.

Lurid Bolete is rather common on clay soils in warm parks, churchyards and cemeteries during late summer and early autumn. It grows with deciduous trees, e.g. beech, lime, birch and oak.

OTHER SIMILAR FUNGI:
To guard against confusion with the **† Devil's Bolete** and **† Bilious Bolete**, check that your Lurid Bolete has a predominantly brown cap and that the surface where the tube layer is attached to the flesh is red. The two poisonous species lack this red surface.

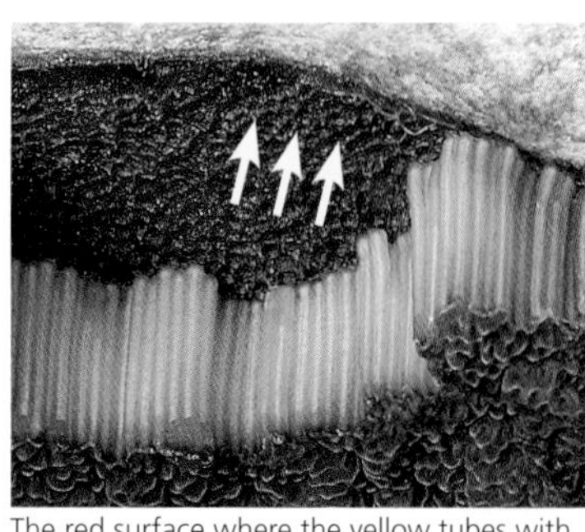

The red surface where the yellow tubes with red mouths adhere to the cap – in section it is seen as a thin, dark line

In Devil's Bolete there is no red colour where the tubes adhere to the cap

Sectioned Devil's Bolete – there is no dark line between the tubes and the flesh of the cap

† Devil's Bolete (*Rubroboletus satanas*) is a large bolete with a pale grey to beige cap, red tube mouths and a thick, club-shaped stem. The net ornamentation of the stem is yellow at the top but lower down becomes a very characteristic rose-pink colour. The surface where the tubes adhere to the cap is yellow and the flesh slowly becomes moderately blue when cut. The taste is mild, but the smell is strong, sweet and rather nauseous and is often likened to pigs dung. The smell is usually strongest from the surface of the cap of old fruitbodies.

Devil's Bolete is rare. It grows in warm places on rich soils with beech, oak and lime, typically in parks, gardens and open woodland in the southern parts of Britain and Ireland. **The species is responsible for some very unpleasant poisonings, causing severe vomiting and diarrhoea**.

† Bilious Bolete (*Rubroboletus legaliae*) is also a large bolete with orange net ornamentation on the stem. The cap is pale without much brown, but usually has obvious pinkish shades, especially near the margin. The surface where the tubes adhere to the cap is yellow. The flesh slowly becomes moderately blue and it lacks the unpleasant odour of Devil's Bolete.

Bilious Bolete is very rare. It is found in warm places on clay soils with beech, for example. **It is believed to cause poisoning, with vomiting and diarrhoea**.

The boletes in the genus **Leccinum** *are recognizable by their long stems covered in fine scales. In most species, the scales are dark and easy to see. All species are considered edible after thorough cooking, although some people get an upset stomach from the orange species (see opposite).*

OTHER SIMILAR FUNGI:
There are no similar boletes with long, scaly stems.

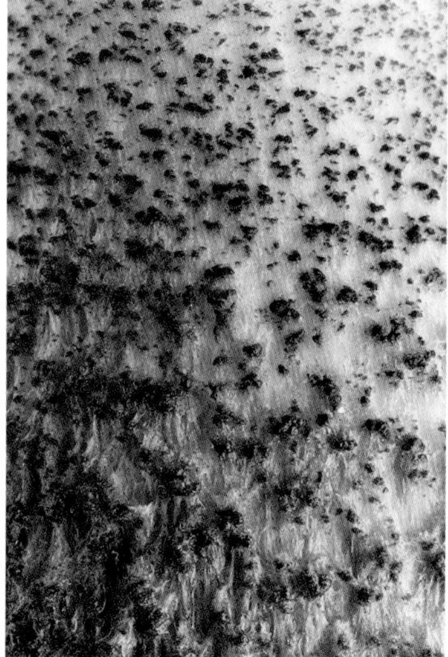

Stem surface with small, dark scales

• **Brown Birch Bolete** (*Leccinum scabrum*) is the most common species in the genus. It has a brown cap and its flesh is either unchanged or becomes slightly orange-red when cut.

The species of *Leccinum* are usually tightly linked to specific host trees. Brown Birch Bolete, for example, always grows with birch.

There are six very similar brownish *Leccinum* species, many of which turn bluish or greenish in the stem base when bruised or cut.

Brown Birch Bolete and its brown-capped relatives are all quite soft-fleshed and so are not very good edible fungi. However, young specimens can be used as a supplement to other mushrooms.

• **Orange Birch Bolete** (*Leccinum versipelle*) belongs to the group of orange-capped species whose flesh turns dark grey when cut. It grows with birch and is somewhat rarer than its brown-capped cousin. A very similar species, Orange Bolete (*Leccinum aurantiacum*), is found scattered with oak and poplar trees.

Orange Birch Bolete and its allies have rather firm flesh and a better taste than the brown species of *Leccinum*. Unfortunately, they contain some somewhat toxic substances that cause a stomach upset if eaten raw. After being cooked thoroughly, they are considered to be good edible mushrooms, but still some people report stomach problems. Therefore, start out with caution, so that you can detect any unwelcome reactions to it early on.

The slimy boletes in the genus **Suillus** *are, when young, completely wrapped in a slimy universal veil. When mature, the cap expands and the slime remains on the cap surface and at base of the stem. Some species also retain a slimy ring. All grow with conifers and are edible.*

OTHER SIMILAR FUNGI:
– slimy boletes with a ring can hardly be confused with other fungi.
– slimy boletes without a ring: see under **Velvet Bolete** opposite.

The slimy boletes are quite soft-fleshed, and this, together with their sliminess, makes them less appetizing. Therefore, it is especially important to collect very young, firm fruitbodies and pull the slime layer from the cap with a knife (see the technique under Slimy Spike (*Gomphidius glutinosus*), page 143).

Young fruitbodies of Slippery Jack

• **Slippery Jack** (*Suillus luteus*) is the best edible species of the genus *Suillus*. It forms medium-sized fruitbodies with a dark brown, very slimy cap, pale yellow tubes and a large, slimy ring on the stem.

Slippery Jack grows with pine trees. It is widespread and common throughout the autumn.

On alkaline soil under pine, the rather identical and edible Weeping Bolete (*Suillus granulatus*) is found. This, however, completely lacks the ring.

• **Larch Bolete** (*Suillus grevillei*) forms medium-sized fruitbodies with orange-brown caps and a small, slimy ring. It always grows with larch and is one of the most common boletes in gardens. It fruits all autumn.

With somewhat softer flesh than Slippery Jack, this species certainly does not rank highly among the edible mushrooms. However, young specimens that have had their slime removed (see page 143) may be used as a supplement to other mushrooms.

• **Velvet Bolete** (*Suillus variegatus*) is a species of *Suillus* without a ring. The cap is initially felted when young, becoming scaly and covered with only a very thin layer of slime. The tube mouths have a characteristic greyish-yellow colour.

Velvet Bolete is a moderately good edible fungus found with pine trees throughout autumn.

Other similar fungi:
– **Bay Bolete** may be somewhat slimy, but its tube mouths quickly turn blue when bruised, p. 74.
– **Peppery Bolete** has orange-brown tubes, yellow flesh and tastes peppery, see below.

• **Peppery Bolete** (*Chalciporus piperatus*) resembles a *Suillus* but has large, brownish-orange tubes, yellow flesh and a slightly hot, peppery taste. It is not poisonous, but is generally not considered edible due to the taste.

It fruits from August to October and is often found in spruce forests though it is probably not connected to the trees but is rather a parasite of the mycelium of Fly Agaric.

Sectioned Peppery Bolete

***Tooth fungi** have a stem and a cap, but unlike many such mushrooms the underside of the cap is covered with small teeth or spines. The edible species belong to the genus of hedgehog mushrooms (*Hydnum*). These are all soft-fleshed and their caps are whitish, light yellow-orange or brownish-orange. They form mycorrhiza and grow on the ground in woodlands.*

OTHER SIMILAR FUNGI:
There are a number of other genera with teeth or spines underneath, see opposite. All these are inedible, but none are poisonous. If you stick to the soft-fleshed species with whitish, yellow-orange or brownish-orange colours, you are safe.

•• **Wood Hedgehog** (*Hydnum repandum*) forms rather large, soft-fleshed fruitbodies. The cap colour is difficult to describe, but as the pictures show, the margin can be almost white, while the middle is often pinkish-yellow to yellowish-orange. The flesh and surface often become darker orange-brown when bruised and it may have a sour or fishy smell. The fruitbodies may even emit almost urine-like odours during frying but do not despair: well-done Wood Hedgehog is an excellent edible mushroom.

This species forms mycorrhizal associations with deciduous trees, and fruits from August to October.

Wood Hedgehog is actually a complex of more than ten very similar but poorly understood species. It is thus not easy to give an exact description of its habitat. Some types are found in beech woods on clays, while others can be very common on poor soils with beech or conifers.

There are also more slender, deep brown-orange types within the Wood Hedgehog species complex. These are just as good edible mushrooms as the pale and more fleshy versions.

• **Scaly Tooth** (*Sarcodon squamosus*) and other species of *Sarcodon* usually are brown and are often very scaly. They have rather soft flesh and as such look inviting, but they often have a bitter and unpleasant taste. Within this group there are several very similar species which are difficult to tell apart; some are rare and so all should be left to adorn their habitats rather than be collected.

Scaly Tooth is soft-fleshed and often has a bitter taste

• **Orange Tooth** (*Hydnellum aurantiacum*) and other species of *Hydnellum* have brown, orange or bluish colours and are often rather tough. There are many rare species which are difficult to recognize. None are known to be poisonous, but Devil's Tooth (*Hydnellum peckii*) has a very unpleasant bitter taste. All are inedible because of their toughness.

Orange Tooth is tough

chanterelles

***Chanterelles** are funnel-shaped or cone-shaped with a solid or hollow stem. The outside or underside of the cap shows irregularly branched veins. Chanterelles grow on the ground and form mycorrhizal associations with trees; all are edible.*

The chanterelles are among the best edible mushrooms. Where many mushrooms are said to taste more-or-less of 'mushroom', the chanterelles have their own, very characteristic, taste of – well – chanterelle. In addition, they have a firm consistency. Many mushrooms become soft or even slimy during cooking, but the chanterelles retain a pleasant crispiness when fried. A simple piece of toast with butter and stewed yellow chanterelles is the ultimate mushroom treat.

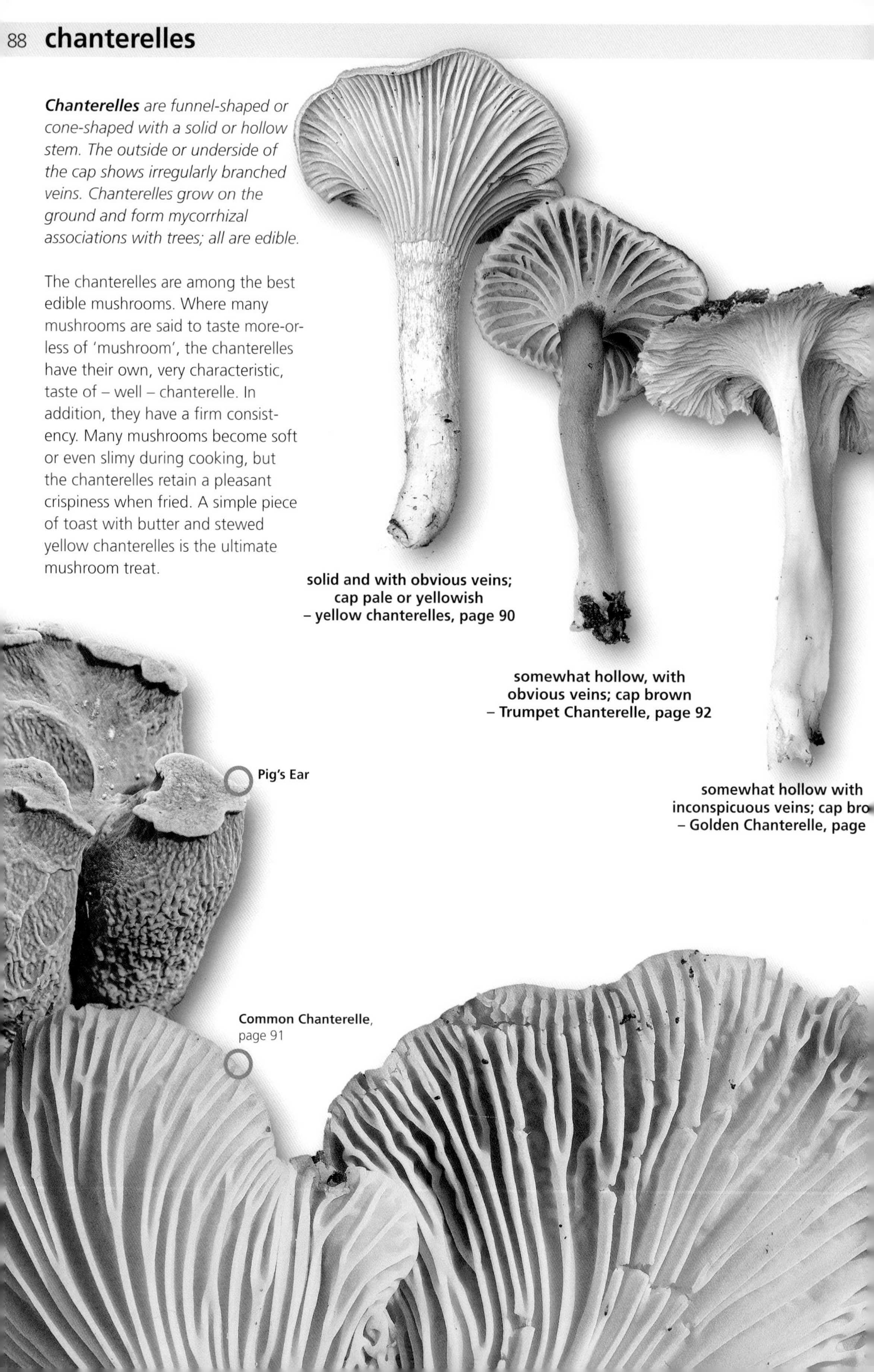

solid and with obvious veins; cap pale or yellowish – yellow chanterelles, page 90

somewhat hollow, with obvious veins; cap brown – Trumpet Chanterelle, page 92

somewhat hollow with inconspicuous veins; cap bro – Golden Chanterelle, page

Pig's Ear

Common Chanterelle, page 91

OTHER SIMILAR GROUPS OF FUNGI:

– **False Chanterelle** (*Hygrophoropsis aurantiaca*) is rather similar to the yellow chanterelles, but it has true gills which are regularly forked, see below and page 91.

– **funnels** may be shaped like chanterelles, but have whitish true gills, see to the right. **Some species are poisonous**.

– **rollrims** are funnel-shaped, but have normal, brown gills that darken when touched, see to the right. **They are poisonous**.

– **Pig's Ear** (*Gomphus clavatus*) is large and fleshy with a chanterelle shape and branched veins on the underside. When young, it is violet, but with age it becomes more brownish (see opposite). It is a good edible fungus, but is extremely rare (perhaps extinct) in Britain and declining in many parts of continental Europe.

•† **Funnels** (*Clitocybe* and *Infundibulicybe*) have funnel-shaped fruitbodies and range from small to large in size. They have normal, unbranched, whitish gills.

Species of Funnel are found both in woods and on open land. **Some are edible, but certain small, whitish funnels that grow on grassland are very poisonous**.

funnels

† **Rollrims** (*Paxillus*) form rather large, brown fruitbodies with the edge of the cap rolled under, a finely downy edge and decurrent gills. When bruised, the pale brown gills change to dark brown.

Some species of Rollrim grow on poor soil with conifers while others prefer clay soils with, for example, lime.

Rollrims have a unique poisonous effect. They are initially edible, but when you have eaten them on several occasions, you may develop an allergic reaction that causes the body's immune system to attack your own red blood cells – in extreme cases with fatal consequences.

rollrims

False Chanterelle, page 91

●● ***The yellow chanterelles*** *form solid, apricot-yellow fruitbodies with irregularly branched veins that run from the cap margin and down the stem. They form mycorrhizal associations and grow on the ground.*

The fruitbodies are usually rather small, 1–8 cm in diameter. The aroma is pleasant and sweet, but the raw flesh tastes peppery (their German name is 'Pfifferlinge'), and **cannot be eaten raw**.

Yellow chanterelles are among our very best edible mushrooms. They have a strong taste and are a treat when stewed and served on toast.

OTHER SIMILAR FUNGI:
– the most obvious candidate for misidentification is **False Chanterelle**, which is soft-fleshed and has true gills with a regularly forked branching pattern, see opposite.
– for more general confusion species, see pages 88–89.

The underside of the Common Chanterelle is covered by branched veins

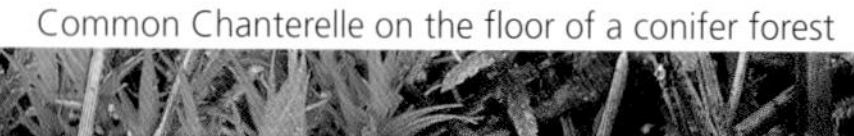

Common Chanterelle on the floor of a conifer forest

Common Chanterelles in abundance

•• **Common Chanterelle** (*Cantharellus cibarius*) forms mycorrhizal associations with a wide variety of trees, but seems especially fond of spruce, pine and beech on rather poor soils.

Search for Common Chanterelles during summer and throughout the autumn, for example, alongside roads and tracks in spruce or pine plantations. With luck, you can, like the boys in the picture, find 'gold mine' after 'gold mine'.

•• **Pale Chanterelle** (*Cantharellus pallens*) is similar to Common Chanterelle, but the fruitbodies are thicker, larger and more fleshy. The colour is also somewhat paler, although it turns brown-orange where it is bruised by picking.

Pale Chanterelle is not common. It prefers clay soil; look especially in forests with slopes, for example, where a stream has cut through the landscape. It has a very fine taste and structure and is one of my favourite edible fungi.

• **False Chanterelle** (*Hygrophoropsis aurantiaca*) looks like a thin-fleshed and very orange Common Chanterelle, but is actually more closely related to the boletes. The underside of the fruitbodies is lined with regularly forked gills (see comparison on pages 88–89).

False Chanterelle grows as a decomposer on the ground and on wood. It fruits during autumn in coniferous forests. It is generally considered edible but may cause stomach upsets. In any case, the soft, rather tasteless flesh makes it quite uninteresting as an edible species.

• ***The brown chanterelles*** *are thin-fleshed and have a hollow stem and brownish cap colours. The underside is covered by shallow, branched wrinkles or veins.*

OTHER SIMILAR FUNGI:
– **Jellybaby** is rubbery or jelly-like and has a head-shaped top instead of a cap, see opposite.
– for more general confusion species, see page 88–89.

The brown chanterelles are excellent edible mushrooms. They may not be quite on a par with their yellow cousins, but they can often be found in abundance so they can easily fill a pie or a pizza.

Because the brown chanterelles are rather thin-fleshed, they are also very suitable for drying.

• **Trumpet Chanterelle** (*Craterellus tubaeformis*) has well-developed veins underneath the cap.

It is found on poor soil with mycorrhiza-forming trees. It can, for example, grow in large quantities in young spruce plantations, but also in beech woods. Where the soil is clayey, search where the otherwise nutrient-rich soil is covered by a thin, acidic, poorly decomposed humus layer.

It typically fruits rather late, from September to November.

Trumpet Chanterelle in coniferous forest

• **Golden Chanterelle** (*Craterellus lutescens*) is very similar to Trumpet Chanterelle, but has only very shallow wrinkles under the cap. It is a fine edible fungus, but is very rare in Britain and Ireland. If, on the other hand, you travel in Norway or Sweden, there is good chance of finding the species. Unlike Trumpet Chanterelle, Golden Chanterelle prefers calcareous soils. It forms mycorrhizal associations with conifers.

• **Jellybaby** (*Leotia lubrica*) can be confused with young trumpet chanterelles. Its fruitbodies consist of a yellow, somewhat slimy stem with a greyer, rather globular head on top. The whole fruitbody has a rubbery consistency.

Although Jellybaby is not poisonous, it has no value as an edible fungus. It often grows with beech, so if you collect your Trumpet Chanterelles in deciduous woods you may well find a Jellybaby in amongst them.

While chanterelles are basidiomycetes, Jellybaby belongs to the other major group of fungi: the Ascomycota.

Horn of Plenty

•• **Horn of Plenty** (*Craterellus cornucopioides*) is unique with its hollow, funnel-shaped, black fruitbodies. The size is very variable, from a few and up to 15 centimetres tall in old, well-grown specimens.

Other similar fungi:
– the only other large, black, hollow fungi found in beech woods are the old, dried-up fruitbodies of **Blackening Brittlegills**, and these are hardly appetizing! See opposite.

Horn of Plenty prefers nutrient-rich, clay soil, where it forms mycorrhizal associations with beech. It is rather common in old beech forests, often together with Trumpet Chanterelle (*Craterellus tubaeformis*) and Pale Chanterelle (*Cantharellus pallens*).

Despite its slightly sinister appearance, the Horn of Plenty is an excellent edible mushroom with a strong taste and a good consistency. Its rather gloomy appearance is reflected in its German name 'Totentrompete' ('Death Trumpet'). Since the fruitbodies are hollow all the way to the bottom, you must remember to split them all the way through to remove woodlice and small snails before they end up in the pan. Horn of Plenty is good fried, but it also works well in sauces and pies. As it is thin-fleshed, it is also very suitable for drying, so you may still have plenty of mushrooms for a cold winter's day!

• **Blackening Brittlegill** (*Russula nigricans*) forms large fruitbodies with stem, cap and gills. These are at first grey-brown, but as time goes on, they do not rot away like other soft-fleshed fungi, but are instead mummified into black 'corpses' that remain in the forest well into winter. It is in this state the hungry mushroom hunter can briefly confuse them with Horn of Plenty.

Blackening Brittlegill forms mycorrhizal associations with beech and grows in the same places as Horn of Plenty, but it is inedible. For more about brittlegills see the following pages.

***Brittlegills** are mushrooms with crisp flesh largely lacking longitudinal fibres. They have no kind of veil, and therefore never have a ring on the stem. They have white or yellowish spores and form mycorrhizal associations with trees. No European species are poisonous, but many taste hot.*

Other similar groups of fungi:
– the poisonous **††amanitas** have a veil that remains as a ring on the fibrous stem. When young, they are totally covered in a universal veil; see more about veils and amanitas on page 128.
– **milkcaps** also have crisp flesh, but in addition have milky juice, see page 106.
– young fruitbodies of **† rollrims** may have crispy flesh. Rollrims are recognizable by their light brown gills that darken when touched, see page 89.

Brittlegills (and milkcaps) have crisp flesh

Typical agarics have fibrous flesh

There are almost 200 species of brittlegills in the UK. They are very difficult to separate, but fortunately there are no poisonous species known from Europe. There are, however, deadly poisonous brittlegills in east Asia.

Some species are hot and can burn as badly as the strongest chilli; these are absolutely unsuitable as edible mushrooms. Some others are mild but rather uninteresting, while still others taste excellent.

To identify brittlegills you must know if they are mild or hot. Start by simply letting the mushroom touch your tongue. If nothing happens, then take a small bite of the gills. Always spit out after tasting.

on the following pages the brittlegills are divided by colour:

red brittlegills, page 97

greyish-red brittlegills, page 100

yellow-orange to yellow brittlegills, page 102

green brittlegills, page 104

•• Perhaps the best edible species among the brittlegills is the **Crab Brittlegill** (*Russula xerampelina*) and its close relatives. It is characterized by the red cap and stem and by the fact that the stem turns brown after bruising. The gills are pale yellow. The taste is mild and with age the fruitbodies acquire a strong odour of seafood.

The Crab Brittlegill most often grows with pine trees and less frequently with spruce and fir. It prefers light soils, both acidic and calcareous, and may be found scattered in conifer plantations in September and October. There are several similar species with other hosts.

Young Crab Brittlegills are very delicious fried and stewed with cream.

Other similar fungi:
– other **red brittlegills** and **†Fly Agaric**, see pages 98 and 131.

The carmine-red stem of the Crab Brittlegill becomes brown when bruised

There are several good edible species among the mild-tasting, bright red brittlegills. Crab Brittlegill (see previous page), for example, is an excellent edible mushroom and the Hintapink (below) is also good and can be collected in large quantities during summer. However, some of the most fiery hot brittlegills are also bright red – and then there is of course the Fly Agaric.

OTHER SIMILAR FUNGI:

– †**Fly Agaric** (*Amanita muscaria*) has fibrous stem flesh and a partial veil that is left as a ring around the stem. The young fungus is also wrapped in a white universal veil which is later left as white scales on the cap. Brittlegills, on the other hand, have no veils at all. But beware: the veil remnants are loose and rain can wash them away and leave the cap completely smooth and red! See more on page 131.

– among **other red brittlegills**, the mild species are harmless, while the hot ones are inedible. The most common, hot species in coniferous forests is **The Sickener**, see opposite.

remnants of the universal veil

Fly Agaric

ring

• **Hintapink** (*Russula paludosa*) is a large edible red brittlegill. It forms fruitbodies with whitish to cream gills, a white stem with red flushes and a shiny orange-red to red cap. It has no noticeable smell but often tastes a bit hot when raw.

Hintapink grows with spruce and pine on poor soil. It is rather common in Scotland but rare or absent elsewhere in Britain and Ireland. It typically fruits as early as July and lasts until October.

Despite the slightly hot taste, Hintapink is a good edible mushroom when cooked.

Hintapink in a dry summer forest

• **The Sickener** (*Russula emetica*)) is one of the most common very hot brittlegills. It mainly grows in coniferous forests and can be recognized by its bright red cap contrasting the pure white stem and gills. If in doubt, carefully place the tip of the tongue against the gills of a fresh fruitbody.

Although probably not really poisonous, The Sickener is simply absolutely inedible because of its taste – which technically is not a taste, but a pain (do not rub your eyes by mistake after handling the mushroom!).

The very similar **Beechwood Sickener** (*Russula mariei*) is common in beech woods and is also inedible.

The cap cuticle does not quite reach the cap margin

• **The Flirt** (*Russula vesca*), also known as Bare-toothed Brittlegill, is one of our better tasting brittlegills. It forms medium-sized fruitbodies with a rather characteristic, pale greyish-red cap colour and mild taste. The stem and gills are white. As a diagnostic feature, the cap cuticle barely reaches all the way to the edge, so especially older specimens 'show their teeth' along the cap margin (see photo to the right).

Very common during late summer and autumn in beech woods, The Flirt can also grow with other deciduous trees and with conifers.

OTHER SIMILAR FUNGI:

– other mild or hot brittlegills with a grey-red and light purple cap, see opposite. However, none of these show their teeth as The Flirt does.

• **Darkening Brittlegill** (*Russula vinosa*) is a large brittlegill with a shiny, greyish wine-coloured cap, cream gills and white stem. With age, the gills, stem and flesh turn grey. It has a mild taste and is a fine edible mushroom.

Scattered in old coniferous forests and plantations in Scotland, Darkening Brittlegill is not known elsewhere in Britain and Ireland. It is however widespread in Scandinavian pine forests.

The gills, stem and flesh of Copper Brittlegill also become grey with age, but this species has a more orange cap, see page 102.

• **Olive Brittlegill** (*Russula olivacea*) is a very large brittlegill with a dull, wine-coloured cap, which often also has olive green shades. The gills become yellow with age and the stem is white with red tones especially as a red collar at the top. The taste is mild and it is an acceptably edible mushroom.

Olive Brittlegill is a typical beech forest fungus. It prefers clay soils and is common towards the south.

If you find a large, mild-tasting beech forest brittlegill, with a red collar at the top of the stem, it is surely an Olive Brittlegill.

• There are many hot-tasting brittlegills with greyish-red colours. **Fragile Brittlegill** (*Russula fragilis*) is one of the more common. It is easily recognized by its finely serrated gill edge (best seen with a lens) and its habitat on poor soils with oak.

The Fragile Brittlegill is one of many species which taste so strong that you just have to touch the gills with the tip of your tongue to 'burn' yourself.

There are several good edible mushrooms among mild brittlegills with orange and yellow caps (see below), but the most common yellow species, Ochre Brittlegill (see opposite), is unfortunately not one of them.

OTHER SIMILAR FUNGI:
– the deady poisonous **†† Deathcap** (*Amanita phalloides*) has a yellow form resembling this group. It is recognized by having fibrous stem flesh, a volva at the base and a partial veil, often left as a ring hanging around the stem, see more on page 128.
– yellow brittlegills with an hot taste are inedible, see opposite.

Deathcap

the young gills are covered by a veil

the volva (remnants of the universal veil)

• **Copper Brittlegill** (*Russula decolorans*) is a large, orange brittlegill that grows with pine on sandy soil. The gills are cream-coloured and the stem is white. With age, the stem, gills and flesh turn grey, just like the Darkening Brittlegill (page 101).

Copper Brittlegill tastes mild and is a fine edible fungus. It is rather rare in Scotland and absent further south but quite common in Scandinavian pine forests. It fruits from late summer to late autumn.

OTHER SIMILAR FUNGI:
The combination of orange cap, greying stem and growth with pine is unique.

• **Yellow Swamp Brittlegill** (*Russula claroflava*) is a large, bright yellow brittlegill with cream-coloured gills. As in Copper Brittlegill, the stem, gills and flesh turn grey with age, and the taste is mild.

Yellow Swamp Brittlegill grows on peat soil with birch and is widespread in Britain and Ireland. It fruits from late summer to early autumn and is a good edible fungus.

OTHER SIMILAR FUNGI:
– the inedible **Ochre Brittlegill** has paler gills and duller colours, see opposite.
– the very hot **Geranium Brittlegill** is more uniformly beige-yellow on cap, stem and gills, see opposite.

• **Ochre Brittlegill** (*Russula ochroleuca*) is a medium-sized brittlegill with a dull yellow cap, whitish gills and a whitish stem that turns more grey with age. It tastes somewhat hot and has no value as an edible mushroom.

Ochre Brittlegill grows in both deciduous woodlands and coniferous forests. It is one of our most common brittlegills and fruits mostly from mid-September until November.

• **Geranium Brittlegill** (*Russula fellea*) is a rather small brittlegill with pale brownish-yellow fruitbodies. Unlike other yellowish brittlegills, the cap, stem and gills have pretty much the same colour. The taste is burning hot and the mushroom is therefore inedible.

Geranium Brittlegill grows mostly in beech woods. It is common and fruits from mid-September.

There are several good edible mushrooms among the green brittlegills. Many are difficult to identify to species, but all completely green mild-tasting brittlegills are edible.

Remember that brittlegills are naked: they never have a veil, neither around the young fruitbody nor as a cover over the gills and thus never display a volva or a ring. **These are important and distinctive characteristics of poisonous amanitas**.

OTHER SIMILAR FUNGI:

– ††**Deathcap** (*Amanita phalloides*), which is **deadly poisonous**, may resemble green brittlegills. When young it is, however, completely wrapped in a veil that is later left as a volva at the base. It also has a partial veil that covers the young gills. This often remains as fragments hanging from the margin of the cap, or as a ring around the stem, see more on page 128.

– green-capped species of **Roundhead** (*Stropharia*) have a ring on the stem and their gills become grey-black, coloured by maturing spores, see below. They are not poisonous.

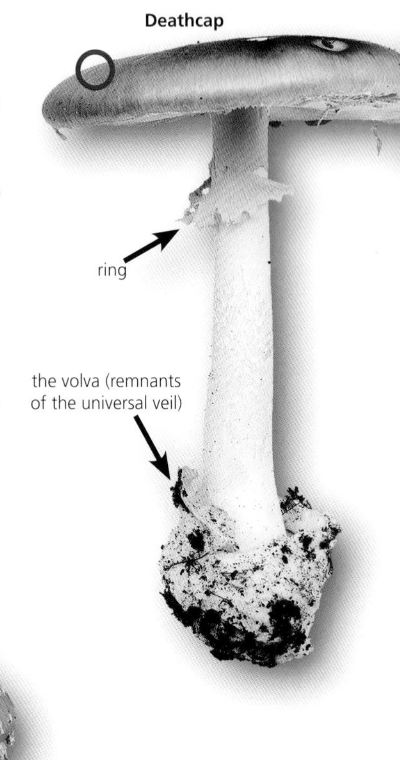

Verdigris Roundhead
(*Stropharia aeruginosa*)

• **Green Brittlegill** (*Russula aeruginea*) has delicate, grass-green caps, whitish gills and a white stem. It tastes mild and is a good edible mushroom.

Green Brittlegill always grows with birch trees. It prefers slightly poor soils in lawns, parks and along roadsides. It is widespread and common from August to October.

• **Greencracked Brittlegill** (*Russula virescens*) is a firm, hard-fleshed brittlegill with a grey-green cap. Upon expansion, the cap cuticle breaks into a chequered pattern. The stem and gills are white and the taste is mild.

Scattered in southerly, warm beech woods on clay soils, this species is able to form fruitbodies under very dry conditions and is therefore often found as early as August.

Greencracked Brittlegill is a good edible fungus.

• **Charcoal Burner** (*Russula cyanoxantha*) can have many colours. Most often, the cap is dominated by light green, but there are usually also areas with more grey and lilac tones. The stem and gills are white. As a rather unique character, the gills of Charcoal Burner are soft and greasy and do not break when rubbed with a finger.

Charcoal Burner grows with beech and is widespread. It can be quite common in August and September.

The mild-tasting Tawny Milkcap (*Lactarius fulvissimus*)

Burgundydrop Bonnet (*Mycena haematopus*)

***Milkcaps** are, put simply, brittlegills with a milk-like juice, most easily seen if you cut the gills with the tip of a knife. The flesh is crisp and without longitudinal fibres.*

The milkcaps often form large, fleshy fruitbodies with more-or-less decurrent gills.

The milk can be white, purple, pink, yellow or clear as water. The colour of the milk is a very important character, so old, dried-out fruitbodies without milk should be avoided.

There are no dangerously poisonous fungi among the milkcaps, but many species taste hot and are completely inedible, and Fenugreek Milkcap (*Lactarius helvus*), which has water-clear milk, may cause stomach upsets (see pages 112 and 113).

the milk can be brightly coloured

From deciduous woods and conifer forests to mountains with willow scrub, milkcaps are everywhere

OTHER SIMILAR FUNGI:
– the only other gilled mushrooms with milky juice are some species of **Bonnet** (*Mycena*, see opposite). Bonnets form small, thin-stemmed fruitbodies, and can hardly be confused with milkcaps. No bonnets with milk are poisonous.

Milkcaps form mycorrhizal associations with trees. They are therefore mostly found in woods, forests and parkland, with both coniferous and deciduous trees. In mountain areas, you may also find milkcaps in low scrub of willow and mountain avens.

If you avoid species with water-clear milk, all other mild-tasting milkcaps can be eaten.

on the following pages the milkcaps are divided into these groups:

brown milkcaps with white milk, page 108 | **milkcaps with orange milk, page 110** | **milkcaps with water-clear milk, page 112** | **milkcaps with hot taste, page 113**

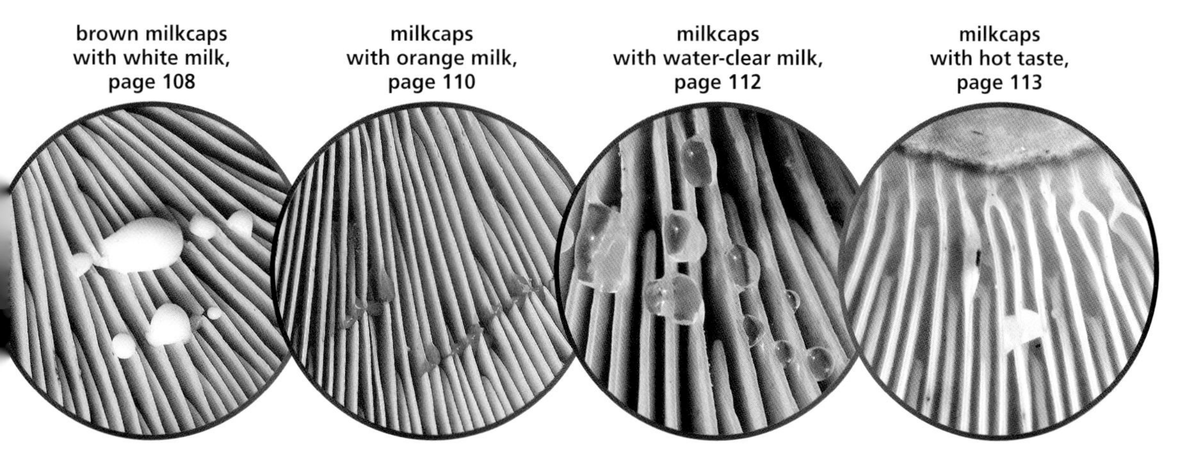

•• **Fishy Milkcap** (*Lactifluus volemus*) forms very compact fruitbodies with a characteristic dull, orange-brown colour on the cap and stem. When cut the gills will secrete an abundance of white milk which becomes brown upon drying. The taste is mild, but often with a slightly bitter aftertaste, and it smells like shellfish.

Other similar fungi:
– it may be confused with other brown, mild-tasting **milkcaps** with white milk, see opposite. These do not smell like shellfish and their milk will not dry to brown spots on the gills.

Fishy Milkcap grows with both conifers and deciduous trees. It is widespread but not common, fruits from August to October and is an excellent edible fungus.

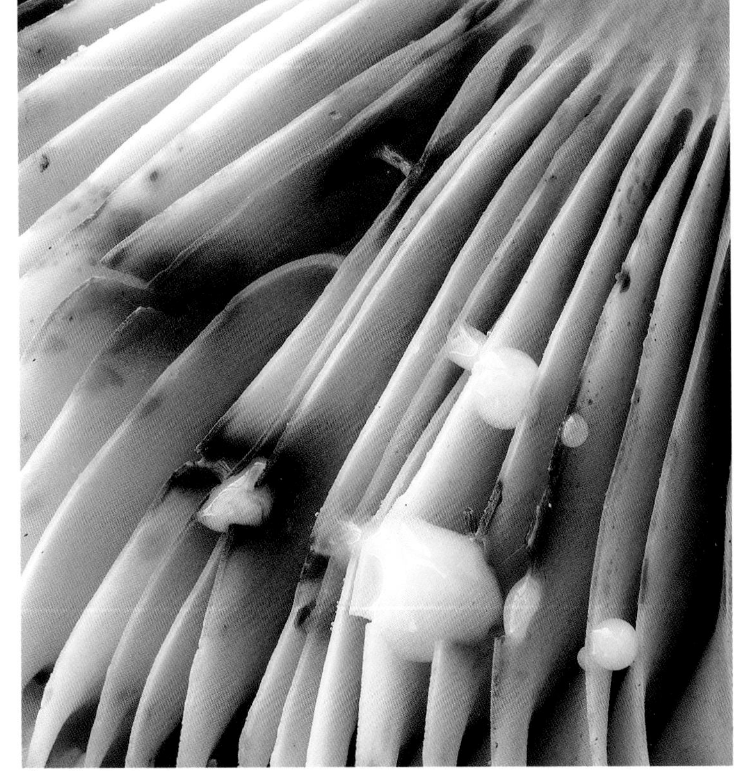

The milk dries into brown spots

• **Oakbug Milkcap** (*Lactarius quietus*) is one of the most common milkcaps. Its dull, grey-brown cap always shows concentric darker zones and has copious white milk. The taste is mild and the smell somewhat unpleasantly spicy (but not fishy).

Oakbug Milkcap grows with oak. It is not poisonous but its unpleasant taste means it is not regarded as edible.

• **Birch Milkcap** (*Lactarius tabidus*) is one among many, small, brown milkcaps with mild, white milk. None of these are considered good edible mushrooms, but they are all quite harmless.

This species grows on moist soil with birch and conifers, such as in bogs. Other brown species are found on drier soils in deciduous and coniferous forests.

• **Rufous Milkcap** (*Lactarius rufus*) is one of the very hot-tasting milkcaps. It forms medium-sized, reddish-brown fruitbodies with a small, pointed peak in the middle of the cap. The milk is white.

This is one of the most common milkcaps in coniferous forests, but because of the taste it is absolutely inedible.

It is among the orange species with orange milk that we find the most important edible milkcaps. The carrot-coloured milk is often quite sparse, and is best seen in a young, fresh specimen. The flesh often turns green some time after bruising.

The species in this group all form mycorrhiza with coniferous trees.

OTHER SIMILAR FUNGI:

– **Sooty Milkcap** (*Lactarius fuliginosus*) and its relatives all have white milk that dries to pink (see page 106, third picture below). They have brown caps, and may have an hot taste. None are poisonous.

After thorough frying the orange-milked milkcaps have a fine consistency and a delicious meat-like taste.

•• **Saffron Milkcap** (*Lactarius deliciosus*) is the best edible fungus in the group. It forms very firm, pale orange fruitbodies and the stems are pitted with small, darker cavities where small drops of juice have gathered and dried. The flesh may slowly turn somewhat green after bruising. There is no particular odour and it tastes mild.

Saffron Milkcap grows with pine and is widespread. It is found in September and October in places where the soil is sandy but not too acidic, for example beside gravel tracks through pine forests. It is a very delicious fungus.

Young fruitbodies excrete small drops which form the cavities in the stem; the broken stem is coloured by orange milk

Freshly harvested Saffron Milkcaps

Saffron Milkcap growing along a gravel road

• **False Saffron Milkcap** (*Lactarius deterrimus*) forms orange fruitbodies with sparse orange milk. The stem is smooth without the pitted surface found on the Saffron Milkcap. With age and after bruising, parts of the cap turn green.

False Saffron Milkcap always grows with spruce and is more common than the Saffron Milkcap. It fruits during September and October. The fruitbodies have a slightly softer consistency than those of Saffron Milkcap, but young, well-fried specimens are still delicious.

† **Fenugreek Milkcap** (*Lactarius helvus*) is the only milkcap that is considered poisonous. It forms large fruitbodies with a dry, somewhat scaly cap surface and brownish colours. The milk is unique by being clear as water. It tastes mild and has a smell that is very reminiscent of curry or stock cubes (in German it is called Maggi-Pilz after a well-known brand).

Fenugreek Milkcap grows with deciduous and coniferous trees on poor, moist soils. It can be used sparingly as a spice for soups, but eating it in larger quantities will result in a stomach upset.

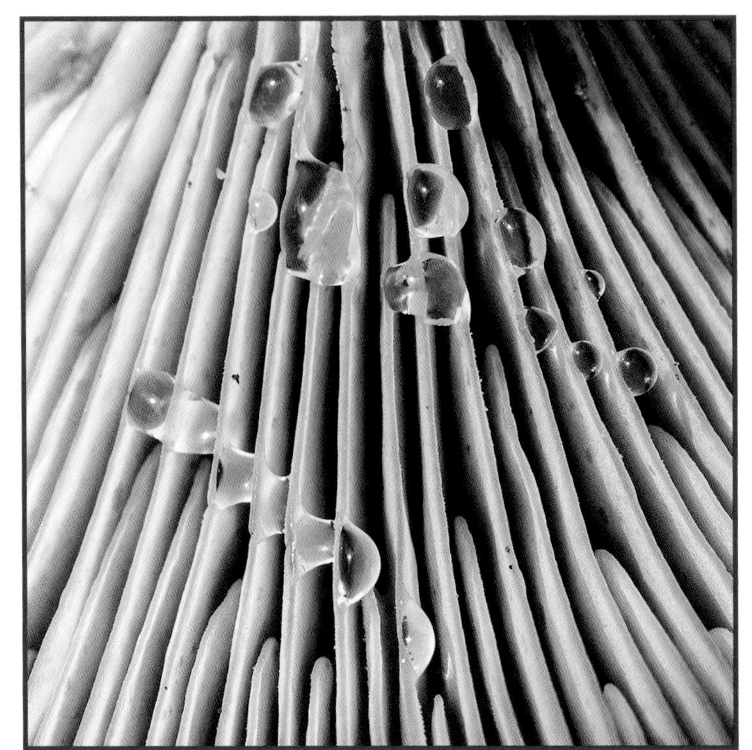

Milkcaps with a very hot taste are not considered poisonous, but are for obvious reasons not really edible either. In some species, however, the hot taste can be removed through salting, see below.

Ugly Milkcap

• **Ugly Milkcap** (*Lactarius necator*) is a large, very hot-tasting milkcap, which grows on nutrient-poor soil with birch and conifers. Its cap is predominantly brownish-green and it has white milk.

Ugly Milkcap is completely inedible because of the taste. The same goes for the super-hot Rufous Milkcap, see page 109.

Bearded Milkcap

•• The species in the **bearded milkcap group** have whitish to pinkish caps with a fringing 'beard' at the cap margin. Their milk is white and very hot.

The species are usually not considered edible, but it is possible to remove the hot taste by salting the mushrooms (see page 41). Salted milkcaps are popular in Eastern Europe. After cooking, they get a strong, slightly meat-like taste.

The two most common species are the pale pink **Woolly Milkcap** (*Lactarius torminosus*) and the paler **Bearded Milkcap** (*Lactarius pubescens*). They both grow with birch, and are widespread and very common throughout autumn.

Woolly Milkcap

*The **agarics** forme a cap with gills beneath. The cap is usually borne on top of a stem. Unlike brittlegills and milkcaps, the stem flesh is structured with longitudinal fibres.*

The agarics are one of the largest fungal groups. The fruitbodies can look very different and vary from a few millimetres to over half a metre in diameter.

The spore colour is an important identification feature and in this book the agarics are divided into white-spored, pink-spored, black-spored and brown-spored species (see the key on pages 48–49).

Unfortunately, spore colours are not always easy to judge. The safest method is to make a spore deposit (see pages 46–47), but you can also learn a lot just by looking at the gills of both young and old fruitbodies. If the gills change colour with age, they are probably coloured by the maturing spores.

The white-spored agarics are the most difficult group to recognize because white spores do not mask other colours effectively. Thus the gills often retain their underlying red, violet, yellow or brown colours. Sometimes, however, one may be lucky to see the whitish spore powder on the surrounding vegetation or other fruitbodies. Otherwise, it is a really good idea to leave a cap on paper for a few hours to make a spore deposit.

The agarics can be further subdivided according to how the gills meet the stem. In some species they run far down the stem (they are decurrent). In others they are attached to the stem over a short length (adnexed) or they swing towards the cap before turning towards the stem (emarginate). And in some species they do not reach the stem at all (free). The different types of gill attachments are shown below.

There are both delicious and dangerously poisonous species among the white-spored agarics. In particular, one must be aware of the poisonous amanitas that are responsible for the vast majority of serious fungal poisonings, see page 128.

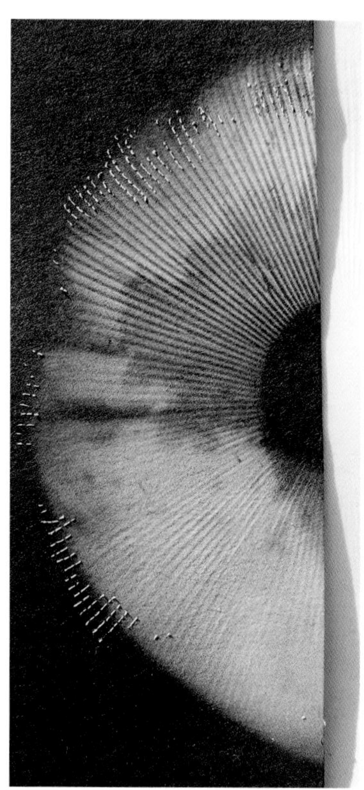

A spore deposit on black paper reveals the white spore deposit of an *Amanita*

on the following pages the agarics with whitish spores are divided like this:

WINTER SPECIES, pages 115–117
SPRING SPECIES, pages 118–119

AUTUMN SPECIES:

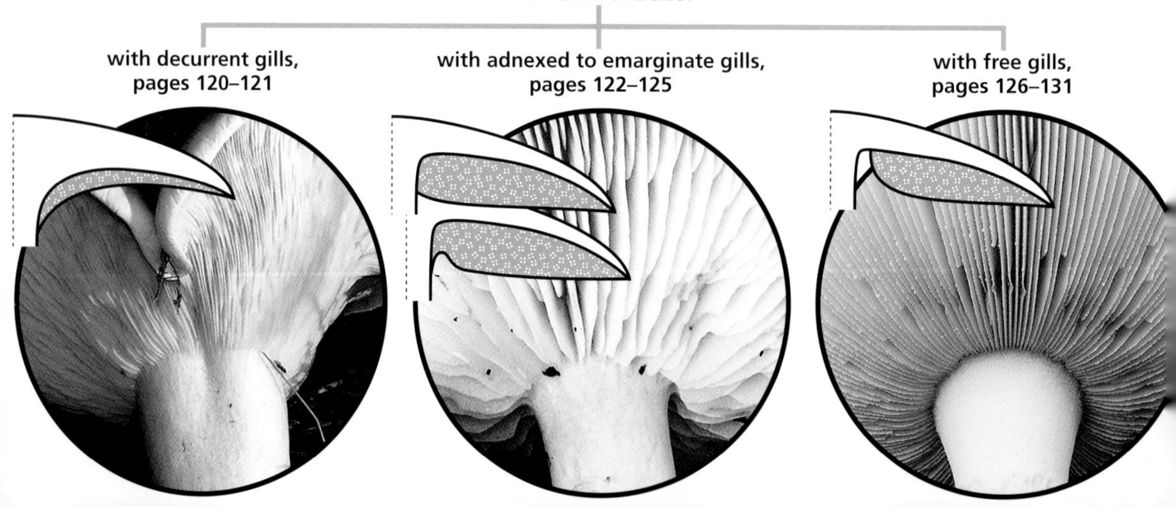

• **Velvet Shank** (*Flammulina velutipes*) is a common winter fungus. Its fruitbodies are small, with a brownish-orange and very slimy cap and whitish gills. The stem has a pale top, but gradually becomes dark with a velvety surface below. The fungus is found on various deciduous trees from November to March.

OTHER SIMILAR FUNGI:
There are several very similar species in the genus, but they are all edible. No other winter fungi look like the shanks.

Velvet Shank is not among the best edible mushrooms due to its slimy cap and mild taste. On the other hand, it is common at a time of year when there is not much else to find. The Enokitake mushrooms for sale in the supermarket are shanks grown in the dark.

adnexed to
emarginate gills

• **Oyster Mushroom** (*Pleurotus ostreatus*) grows on deciduous wood, where it forms large clusters of oblique, greyish-brown to greyish-blue caps. The stem is short and is situated at the edge of the cap, not in the middle as is typical in agarics.

The underside is lined with white gills that continue far down the stem. In young fruitbodies, the gills often branch to form a coarse net down the stem. The fruitbodies are thick-fleshed with a mild taste and no notable smell.

Oyster Mushroom is mostly found from late October to March. It prefers deciduous trees like beech, willow and poplar. Young, fresh fruitbodies are edible with a good consistency but very mild taste.

The gills continue in a net-like pattern towards the base of the stem

OTHER SIMILAR FUNGI:
– **Olive Oysterling** has yellowish colours on gills and stem, and the gills end abruptly instead of running far down the stem, see below. It is not poisonous.
– † **Angel's Wings** is white on the cap, stem and gills, see below. **It is poisonous**.
– there are a number of similar species with smaller, thin-fleshed fruitbodies, white gills and oblique stems growing on wood. Only a few of these are winter fungi and none are poisonous.

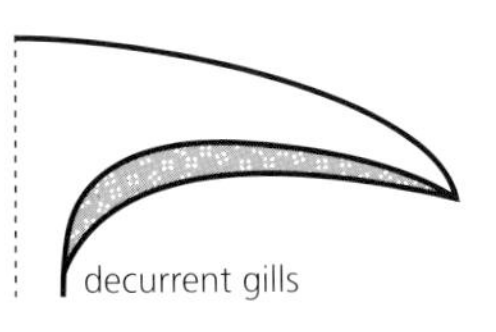

Oyster Mushrooms high in a beech

† **Angel's Wings** (*Pleurocybella porrigens*) resemble slender, thin-fleshed, white Oyster Mushrooms. They grow mostly on conifers and fruit in September and October before the Oyster Mushroom really begins.

If eaten in large quantities, this species is reported as causing severe poisoning. It is mostly found in Scotland, and if you are careful not to eat completely white 'Oyster Mushrooms', you should not be at risk.

• **Olive Oysterling** (*Sarcomyxa serotina*) looks like a small Oyster Mushroom with more yellow and greenish colours. An important character is the gills, which end in an almost straight line at the stem. In Oyster Mushroom the gills continue downwards and form a net-like pattern.

Olive Oysterling grows on deciduous wood and fruits late, typically October to January. It is edible, but not worth the trouble.

• **St George's Mushroom** (*Calocybe gambosa*) forms medium-sized fruitbodies which are usually white or cream-coloured (the caps may turn light brown when very wet). The gills are noticeably crowded together. The whole fruitbody has what mycologists call a farinaceous – or flour-like – taste and smell.

Although described as 'flour-like', the smell of this species has little similarity to that of modern, refined, white flour. The smell is perhaps more reminiscent of cucumber or watermelon – or of old-fashioned, moist flour sacks. A very similar odour is found in Dryad's Saddle (page 57). You really have to experience the smell by smelling a farinaceous fungus to get the idea.

OTHER SIMILAR FUNGI:

– spring species of the **Pinkgill** genus have more distant, less crowded gills which turn pink with age due to their pinkish spores, see opposite. **Some rare species are poisonous**.

St George's Mushrooms grow in lawns and in deciduous and coniferous forests on well-mixed, humus-rich soils. They often form large fairy rings and fruit from April to June.

This is a good edible mushroom, provided that it is fried very thoroughly so the flour-like taste and smell disappear.

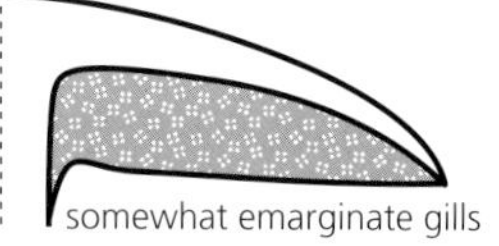

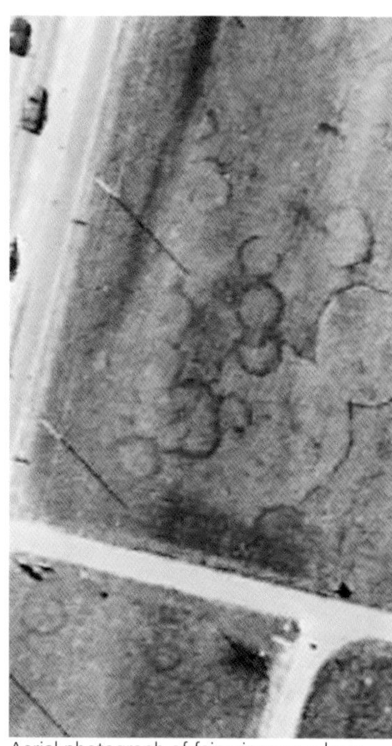
Aerial photograph of fairy rings on a lawn

St George's Mushrooms as they are often found, half-hidden in spring vegetation

† Certain species of **Pinkgill** (*Entoloma*) are poisonous. Some species are small and slender, while others are large and fleshy and can be mistaken for edible mushrooms. The pinkgills have more distant gills (there is more space between them) than St George's Mushroom. Though often whitish or yellowish at first, the gills become coloured by the pink spores as they age. Pinkgills are found both in wooded habitats and open land.

• **Giant Funnel** (*Aspropaxillus giganteus*) forms huge fruitbodies with a somewhat funnel-shaped, whitish cap, whitish, decurrent gills, and a short fibrous, white stem. The caps can be more than 30 cm wide and the stem 1–5 cm thick. A rather distinctive feature is that the gills can easily be detached from the cap flesh (see picture opposite, left). This species grows in woods, gardens and open habitats. It is a parasite that kills and decomposes herbs and grasses.

When Giant Funnel grows in fields, it may form large, conspicuous fairy rings. A 40–50 cm-wide zone appears where the vegetation has been killed by the fungus. Inside this zone, grasses and herbs at first often appear unusually green and lush, due to nutrients that the fungus has released but has not yet absorbed. Because of these zones, the fairy rings are visible even in the absence of fruitbodies, and they can easily be seen on good aerial photographs.

Giant Funnel can cause stomach problems in some people, but for most it is a fine edible fungus.

For safety's sake, start with a small portion, use only young specimens and fry them well so that their strong smell of bitter almonds disappears.

The gills can be detached from the cap

OTHER SIMILAR FUNGI:
– some small, white **funnels** (*Clitocybe*) with stem thicknesses of less than 1 cm can be very poisonous.
– there are funnel-shaped **brittlegills** and **milkcaps** (*Russula* and *Lactarius*) but these have brittle (not fibrous) flesh and the milkcaps also have a milky juice. They are not poisonous but may be very hot. They form mycorrhiza and grow near trees, usually in woodland habitats.
– † **rollrims** (*Paxillus*) have brown colours, and their gills become dark brown when bruised. As with Giant Funnel, Rollrim gills can be easily detached from the cap flesh. Rollrims are poisonous (see page 89). They form mycorrhizal associations and thus grow close to appropriate trees.

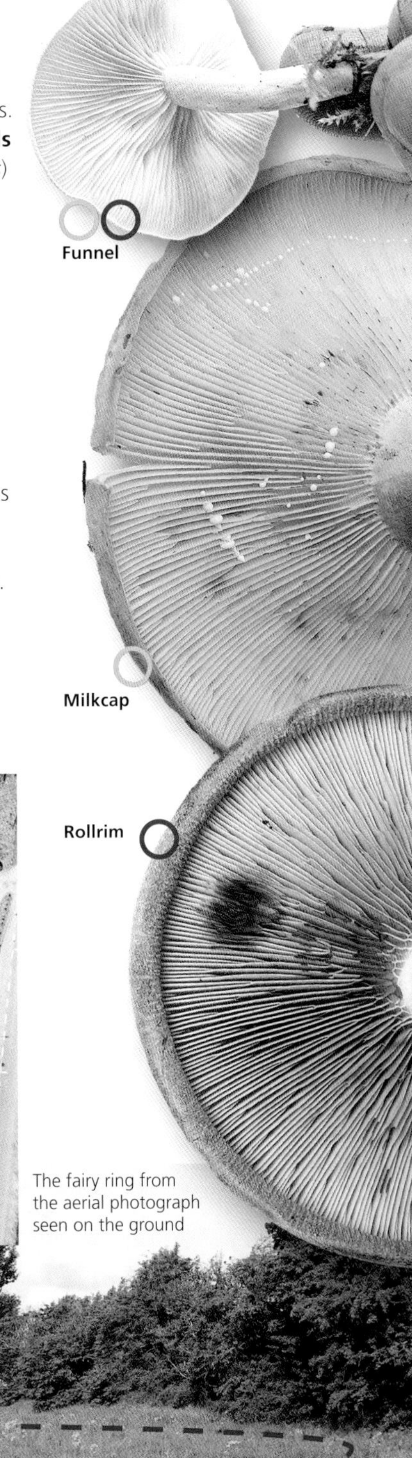

Fairy rings on an aerial photograph. The large one crossing the road is a Giant Funnel

The fairy ring from the aerial photograph seen on the ground

• ***Honey Fungus*** (Armillaria) *is a genus of brownish, white-spored fungi with gills. They are parasites on trees.*

Honey fungi have medium-sized caps with small honey-yellow to brown scales at the centre. The gills, which start white, are first covered by a thread-like or cotton-like veil. With age, the gills become more brownish. They are at first emarginate, but may later become almost decurrent. The veil is left as whitish fluff on the lower stem and as a sometimes flimsy ring.

OTHER SIMILAR FUNGI:
– **scalycaps** (e.g. **Shaggy Scalycap**, see opposite) are rather similar but have brown spores. Young honey fungi have white gills, whereas young shaggy scalycaps have yellowish ones.

Honey fungi live as parasites on deciduous and coniferous trees. They kill the trees and decompose the wood, and do great damage in forests and gardens. The fruitbodies are formed during late September and October.

There are several species of honey fungi and they are very difficult to separate. They are generally edible **but will cause stomach pain and vomiting if eaten raw.**

Honey fungi are considered good culinary mushrooms in some countries, but there are also reports that they may sometimes cause stomach problems. These are probably due to consumption of incompletely cooked fruitbodies in combination with alcohol. It is claimed that fruitbodies growing on coniferous trees are more difficult to digest than those on deciduous trees (see the discussion under Dark Honey Fungus opposite).

If you want to eat honey fungi, you should go for very young, still almost closed caps, see the picture below. Make sure to fry them thoroughly, and start for safety's sake with a small portion to make sure your stomach will tolerate them.

Mature cap with small, dark scales

Freshly picked, perfect, young caps of honey fungi

White, slightly emarginate gills and a ring

• **Bulbous Honey Fungus** (*Armillaria lutea*) has a fluffy, cottonwool-like ring and young specimens usually have yellowish scales on the cap, ring or stem.

The species is common on the wood and roots of deciduous trees and is found rather late in autumn. The young caps are good edible mushrooms.

(•) **Dark Honey Fungus** (*Armillaria ostoyae*) has a more robust ring and has dark, not yellow, scales on the cap. It mostly grows on coniferous wood and fruits in late autumn.

Dark Honey Fungus may be more difficult to digest than other species of Honey Fungus. Therefore, you might avoid specimens collected in coniferous forests or at least test your tolerance with a very small amount at first – and do not serve them to friends without warning!

† **Shaggy Scalycap** (*Pholiota squarrosa*), with its brown colours and pointed scales, may be taken for a Honey Fungus. However, its young gills are pale yellow and darken with age as its brown spores mature. Honey fungi, in contrast, have whitish gills and white spores.

Shaggy Scalycap is common. It grows at the base of deciduous trees and fruits rather late in the year. It is considered to be slightly poisonous.

• **Fairy Ring Champignon** (*Marasmius oreades*) forms fairy rings with numerous small, beige-coloured fruitbodies. They have adnexed, distant gills and a characteristic smell of bitter almonds.

OTHER SIMILAR FUNGI:
There are lots of small fungi in lawns and fields, but none grow in fairy rings and have the combination of such pale brown colour, distant gills and the smell of bitter almonds.

Fairy Ring Champignon is one of the most common species in lawns. Its fairy rings often show a zone of dead grass and are therefore visible all year round. It is a reasonable, albeit small, edible mushroom, but **it must be thoroughly fried before being eaten**.

• **Garlic Parachute** (*Mycetinis alliaceus*) smells strongly of garlic and grows on wood in beech forest.

It has almost the same cap size and colour as the Fairy Ring Champignon, but the stem is long, felted and dark brown. It grows on twigs and stems on the woodland floor and is a typical beech wood fungus. The strong garlic-like smell emerges from both the fruitbodies and its mycelium in the rotting wood.

OTHER SIMILAR FUNGI:
There are no other woodland fungi with a long, black-brown stem and a strong smell of garlic.

Common in old, southern beech woods, Garlic Parachute can be found from August to October. Due to its insubstantial size and strong smell and taste, Garlic Parachute is not a typical edible mushroom, but it can be used as a spice if you want a flavour of garlic without the real stuff.

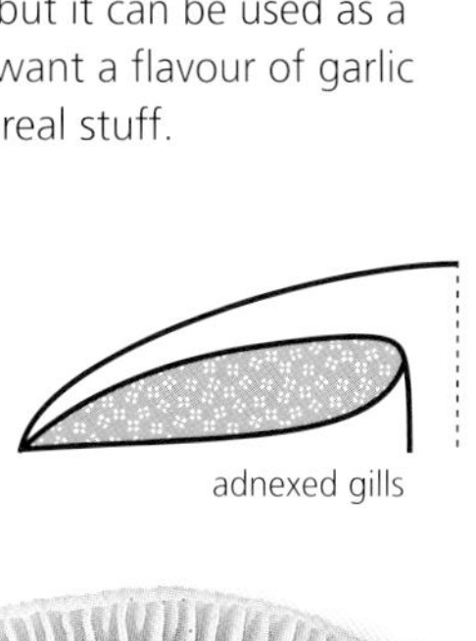

adnexed gills

• **Giant Parasol** (*Macrolepiota procera*) forms very large, stately fruitbodies with scaly caps up to 30 cm wide.

The very young cap is smooth, but as it unfolds it cracks into large scales. In time only the area in the middle of the cap remains whole and brown, while the rest shows the whitish, exposed flesh with scattered, brown scales. The gills are white and free, and the young gills are covered by a membraneceous veil. This later forms a characteristic double ring that hangs loose so that it can be moved up and down the stem without breaking. The surface of the stem cracks into a dark and white snakeskin pattern.

Giant Parasol grows in woods, fields and dunes, and is especially frequent in coastal areas. It is most common in the southern parts of Britain (more scattered in Ireland) and fruits in late summer and early autumn.

If you try to fry a Giant Parasol it will just absorb all the fat you throw at it and not become crisp, so instead coat the whole, flat caps in breadcrumbs. Fried in this way Giant Parasol is a surprisingly good edible fungus.

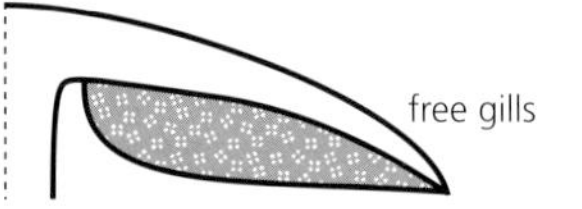

Other similar fungi:

– species of the parasol genus ***Chlorophyllum*** are very similar but generally a bit smaller. They have reddening flesh and a smooth stem surface. Most species of *Chlorophyllum* are edible, but there is also one somewhat poisonous species, which mostly grows in gardens and greenhouses, see opposite.

– some species of **Amanita** have brown and scaled caps, but here the scales are remnants of a universal veil left as small flakes on the cap surface and they can be removed without breaking the cap cuticle. The amanitas have a single ring which is firmly attached to the stem, see opposite.

– other very similar species of **Parasol** (*Macrolepiota*) grow in woods and are also edible.

• Giant Parasol with a complex, double ring

• Brown Parasol with a simple ring

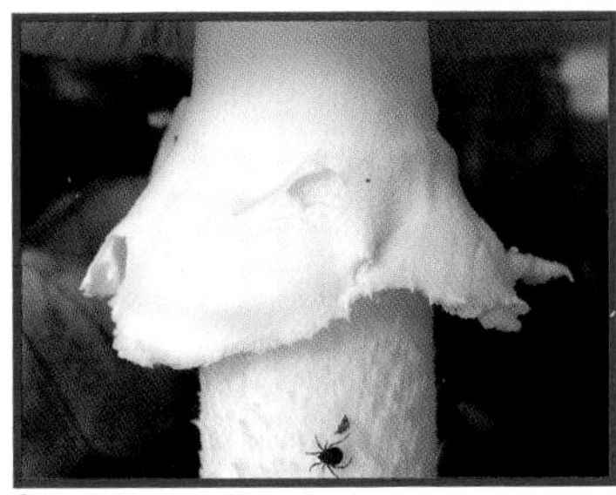
† Panthercap with a simple ring

• **Brown Parasol** (*Chlorophyllum brunneum*) is very similar to a small species of *Macrolepiota*, but has a smooth stem which does not crack into a snakeskin pattern and a simple ring. It grows in gardens and greenhouses and can cause upset stomachs in some people.

There are several other very similar species of the genus *Chlorophyllum* that are edible. These, like the Giant Parasol, have a double, loose ring at the stem.

† **Panthercap** (*Amanita pantherina*) may also look a bit like a parasol, but is smaller and has a non-cracked cap covered with scales, which are the white, loose remnants of the universal veil. It also has a single, hanging ring firmly attached around the white stem.

Panthercap, like all amanitas, forms mycorrhizal associations with trees. It will therefore never grow in open landscape, which is where you will often find the Giant Parasol. **Panthercap contains some neuro-toxins that can cause severe but fortunately not fatal poisonings.**

Any book on edible fungi must include, with appropriate warnings, the most poisonous amanitas. Once you are familiar with Deathcap, Destroying Angel and Fly Agaric, you will feel a lot safer identifying edible species.

Amanitas *(Amanita) are recognizable by their white, adnexed or free gills and by the membraneceous universal veil that surrounds the young fruitbodies.*

The universal veil protects the young fruitbody from insect infestation and dehydration. When the fruitbody matures, the veil bursts, and is left either as a volva around the base of the stem or as white scales on the cap surface.

As the fruitbody expands, the partial veil remains as either a ring around the stem, or fragments hanging from the margin of the cap (or both).

The amanitas form mycorrhizal associations with trees, and are therefore found only near suitable tree species in woods, gardens and parks.

Many amanitas are poisonous, but there are also edible species. The poisonous species contain a wide range of toxins, the most dangerous of which are the amatoxins found in Deathcap and Destroying Angel. The toxins start working long after the mushrooms have been eaten and it is to late to perform stomach pumping. Treatment consists of eating activated charcoal and strong antibiotics, but there are no effective antidotes. The toxins destroy the liver cells and poisonings may end in liver failure and death.

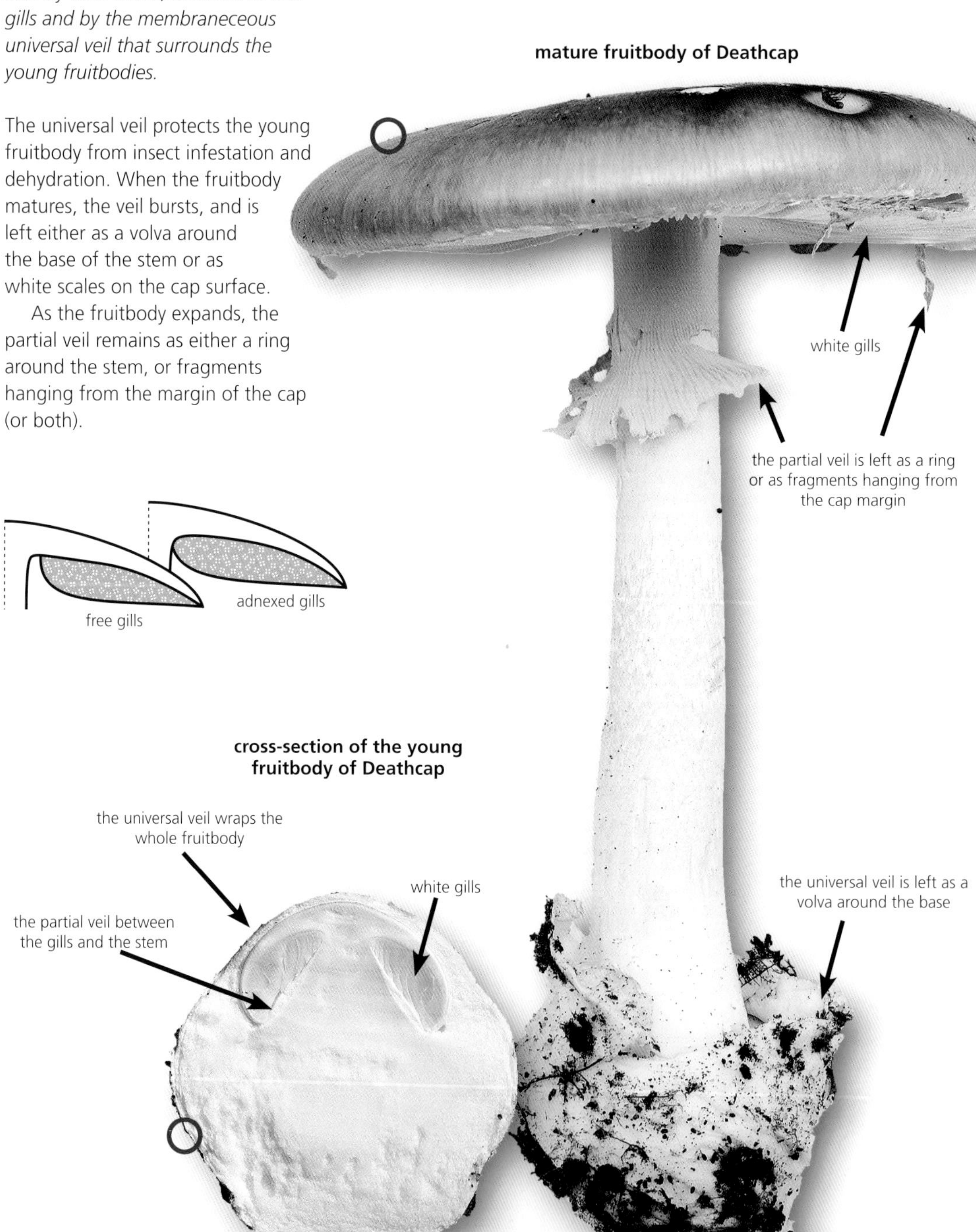

†† **Deathcap** (*Amanita phalloides*) forms medium-sized, beautiful fruitbodies with a pale green, yellowish or pale grey cap. The gills are white and free, and produce white spores. The stem has a somewhat cracked surface and often carries the remnants of the partial veil as a hanging ring. In some mature fruitbodies the ring is confusingly absent: here the veil may instead remain as fragments hanging from the cap margin. At the base of the stem is a bulb, and around this you will see the remains of the universal veil – the volva.

Deathcap is very common in beech and oak woods, especially in southern parts of Britain and Ireland. It fruits mainly during August and September.

Deathcap is deadly poisonous. Just a single fruitbody can be enough to cause liver failure and kill an adult (see opposite). Deathcap is responsible for the vast majority of fatal mushroom poisonings in western Europe.

A young, very pale Deathcap breaking through its universal veil

A very yellow version of Deathcap

†† Destroying Angel (*Amanita virosa*) forms medium-sized, slender, white fruitbodies with white spores. The cap is rather high and usually a little asymmetrical. The gills are free, and the stem long and somewhat frayed. The partial veil is usually left on the stem as a hanging ring. At the base there is a bulb, surrounded by a white volva, the remnants of the universal veil.

Destroying Angel grows mostly on poorer soil, where it forms mycorrhizal associations with, for example, beech, oak, spruce and pine.

Just as poisonous as its green cousin, Destroying Angel contains the same toxins as Deathcap. The reason why the latter causes the most fatalities is probably that it is the more common species in and around urban areas.

OTHER SIMILAR FUNGI:

– Destroying Angel can easily be confused with white species of ***Agaricus***. But where the gills of Destroying Angel (and other amanitas) remain white throughout their lives, the gills of the *Agaricus* species soon turn light grey or pink, and further change to dark chocolate-brown on maturity (page 134).

– **False Deathcap** (*Amanita citrina*) is very similar to the Destroying Angel but has a more symmetrical, rounded cap and smells like raw potato. This species is not poisonous but much too difficult to identify to be considered safe to eat.

a young, sectioned Destroying Angel

a young, sectioned Fly Agaric

† **Fly Agaric** (*Amanita muscaria*) forms large fruitbodies with red or orange caps. Remains of the universal veil are usually left on the cap as the characteristic white, loose scales, but after rain these may wash away and the cap can then become completely naked. The gills are white and free or rarely adnexed, and the stem is white and carries a large, eye-catching ring. Since the universal veil is mainly left on the cap, Fly Agaric has no volva at the base.

Fly Agaric forms mycorrhizal associations with many different trees, but is especially frequent with birch and spruce on nutrient-poor soil. It is common throughout Britain and Ireland and can be found fruiting from September to November.

Other similar fungi:
See the discussion under the **red brittlegills**, page 98.

Fly Agaric is poisonous but hardly deadly. It contains a cocktail of various toxins, but not the very dangerous amatoxins found in Deathcap and Destroying Angel. On the other hand, the fruitbodies contain the substances muscimol and ibotenic acid, which can cause sweating, vomiting and an intoxication that can include hallucinations. As the relationship between the various substances is unpredictable, one should never try to use Fly Agaric as an intoxicant – even though it has historically been used as such by Siberian peoples.

Rather violet Wood Blewits in a coniferous forest

• *The* ***violet blewits*** (Lepista) *form medium-sized fruitbodies with pale brownish-pink spores, emarginate gills, and violet or purple colours on the gills and/or stem. The fruitbodies are naked, without any traces of veil or ring, and have a very characteristic, sweetish odour.*

Blewits are decomposers and grow in both woodlands and open habitats. They very often form large fairy rings.

All blewits are edible, but some people do not like the somewhat perfumed taste and smell. However, they are generally considered to be good and safe edible mushrooms.

OTHER SIMILAR FUNGI:

– violet species of **†Webcap** (*Cortinarius*) are recognizable by their thread-like veil between the stem and cap margin in young fruitbodies and by the rusty brown spores (see opposite). **Some species are poisonous**.

– **Amethyst Deceiver** (*Laccaria amethystina*) is smaller with a thinner stem (3–6 mm) and very distant gills (see opposite). It is not poisonous.

Wood Blewit forming an almost perfect fairy ring

• **Wood Blewit** (*Lepista nuda*) is a medium-sized agaric with violet gills, violet stem and violet to brownish cap (see also the picture on page 2). The cap surface is smooth and greasy and the stem is 5–20 mm thick and has a fine, fluffy pattern.

Wood Blewit is a decomposer which can grow in both fields and woods. It has a preference for woodland habitats with a deep layer of organic material, such as coniferous forests with a thick needle cover, and is also fond of heavily composted gardens. It appears quite late, typically in October, and is very common and a good edible fungus.

• **Field Blewit** (*Lepista saeva*) resembles a large, pale, grey-brown version of Wood Blewit, but only the stem is a lilac shade. It is a salt-tolerant species which is especially characteristic of coastal pastures; it has also spread to roadsides affected by winter salting.

Fruiting in October and November, Field Blewit has a southerly distribution and is a good edible fungus.

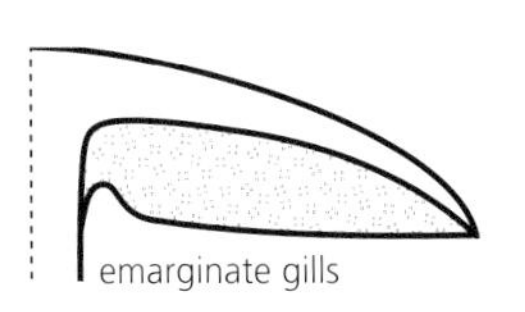

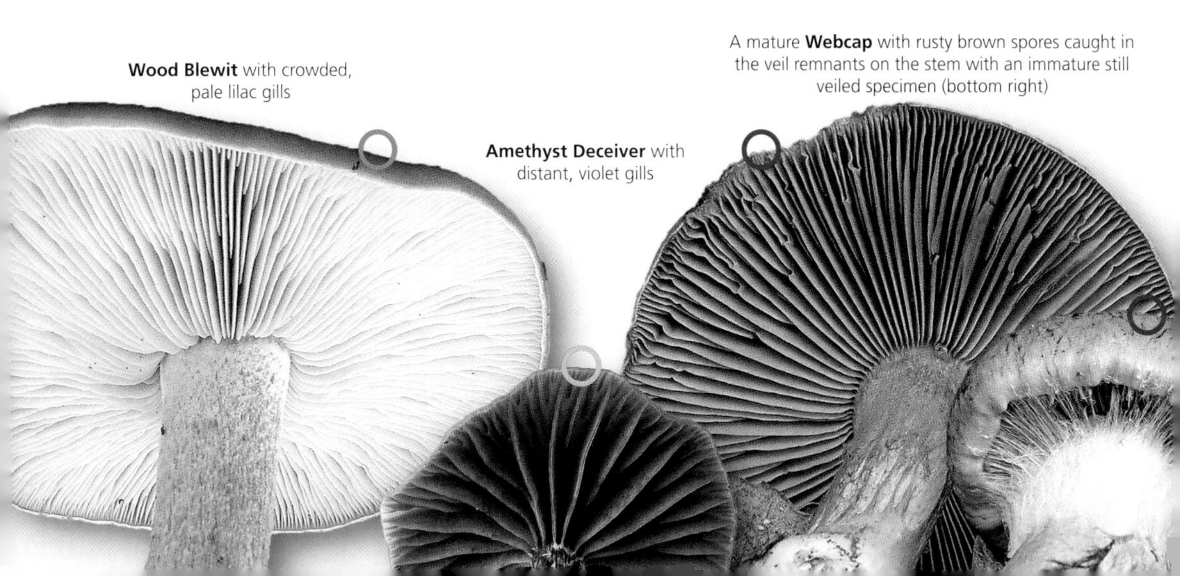

Wood Blewit with crowded, pale lilac gills

Amethyst Deceiver with distant, violet gills

A mature **Webcap** with rusty brown spores caught in the veil remnants on the stem with an immature still veiled specimen (bottom right)

Members of the genus **Agaricus** *have rather fleshy fruitbodies with dark chocolate-brown spores and free gills. They have a partial veil that covers the young gills and later hangs as a ring around the stem. They never have volva at the base of the stem. All species except the Yellow Stainer (*Agaricus xanthodermus *– see page 139) are edible.*

The most important character of th genus *Agaricus* is the colour of the spores. In young fruitbodies the gills are light grey or pink, but as the spores mature, the gills turn chocolate-brown.

OTHER SIMILAR FUNGI:

– the deadly poisonous **††Destroying Angel** has pure white gills and when young is completely wrapped in a universal veil, which remains as a volva around the base, see below and page 130.

– **Yellow Stainer** (*Agaricus xanthodermus*), the *Agaricus* species that can cause stomach upsets, turns butter-yellow when touched and smells of chemicals, see page 139.

– there are many other more fragile fungi with dark spores, such as **brittlestems** (*Psathyrella*) and **fieldcaps** (*Agrocybe*). These have adnexed gills.

Young Field Mushroom and mature Horse Mushroom with a ring and free, dark gills

***Agaricus* species must *never* be eaten unless they show coloured gills**

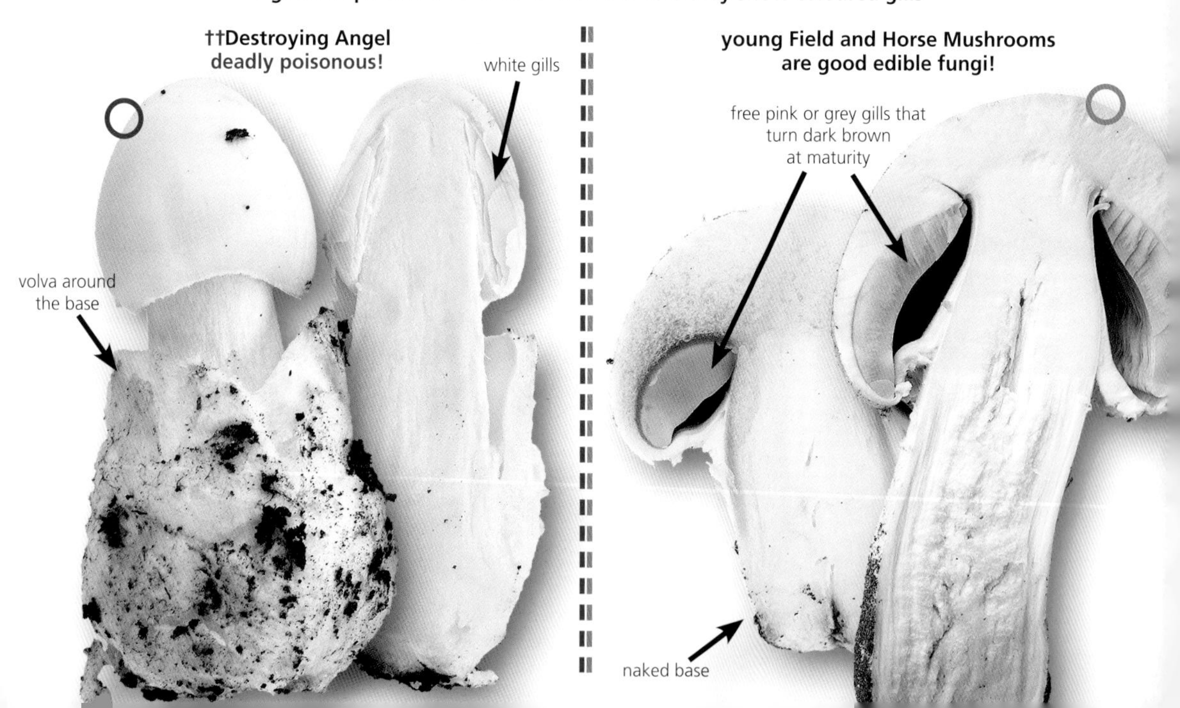

types of *Agaricus* species

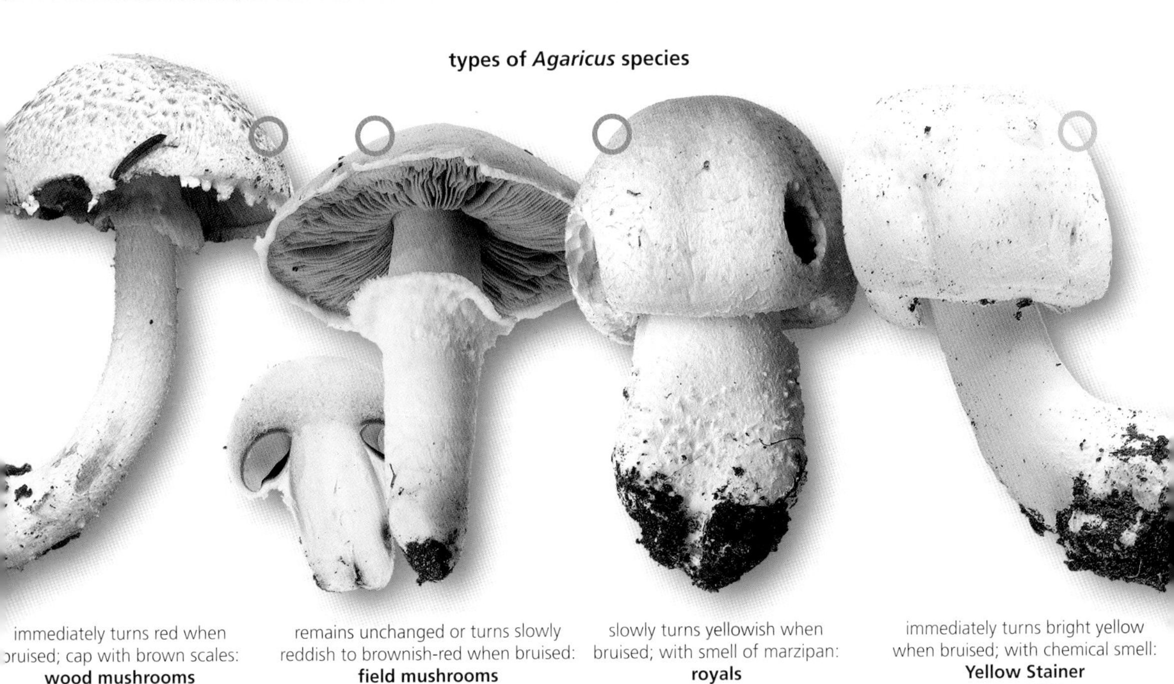

immediately turns red when bruised; cap with brown scales:
wood mushrooms
page 136

remains unchanged or turns slowly reddish to brownish-red when bruised:
field mushrooms
page 137

slowly turns yellowish when bruised; with smell of marzipan:
royals
page 138

immediately turns bright yellow when bruised; with chemical smell:
Yellow Stainer
page 139

Agaricus species typically decompose organic matter in soil or manure. They can therefore live both in woodlands and in the open countryside. There are about 40 species in the UK and many are very difficult to differentiate from each other.

Many species of *Agaricus* form fairy rings. Over the years the rings fragment, so the fruitbodies may grow in arches and rows, rather than complete rings.

Only an arch of fruitbodies is left of this old fairy ring of Field Mushrooms

*The **red stainer** group of* Agaricus *species turn red or brownish-red when cut. In some, this red staining happens quickly and the colour turns deep red (e.g. the **Scaly Wood Mushroom**), while others blush only faintly and slowly (e.g. the **Field Mushroom**).*

OTHER SIMILAR FUNGI:
– some poisonous species of ***Amanita*** look rather similar but their gills remain white, see pages 127 and 130.
– the reddening species of Parasol (*Chlorophyllum*) also have white gills, see page 127.

• **Scaly Wood Mushroom** (*Agaricus langei*) is a medium-sized, brown-capped *Agaricus* which reddens immediately and strongly when bruised. It has brown gills and a well-developed ring with a cogwheel-like underside. It smells pleasant and is a fine edible fungus.

Scaly Wood Mushroom grows in both deciduous woods and coniferous forests and fruits during autumn. There are several similar species, all of which are edible.

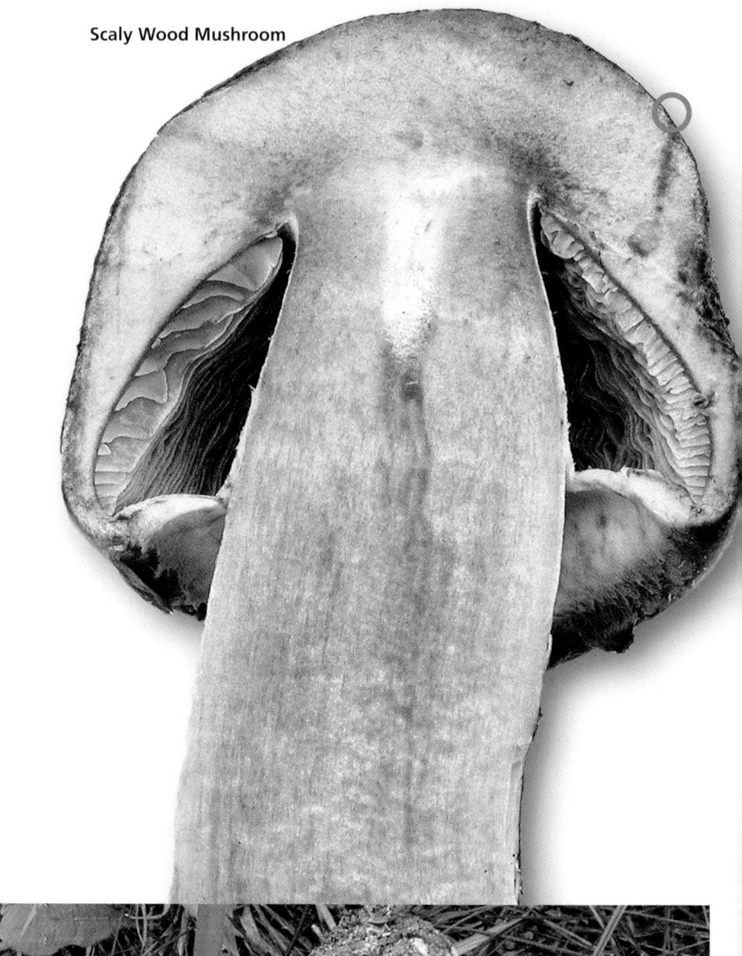
Scaly Wood Mushroom

• **Field Mushroom** (*Agaricus campestris*) forms small to medium-sized fruitbodies with a white cap, pink to chocolate-brown gills and a stem with a small, fragile ring without a cogwheel appearance underneath. The flesh blushes slowly and moderately when cut, and it smells faintly of 'mushroom'.

Field Mushrooms are very common in lawns, parks and grasslands, where they often form fairy rings. They are found during summer and early autumn and are good edible mushrooms. There are many closely related species, all of which are edible.

The simple stem ring of Field Mushroom. Bottom, a fairy ring formed in a lawn

Horse Mushrooms on a lawn; to the right, a fruitbody with yellow patches from bruising in the basket

•• ***Royals*** *are* Agaricus *species with strong marzipan-like smells and which slowly turn yellow when bruised.*

Royals are all excellent edible mushrooms. Unfortunately, they have the ability to concentrate the heavy metal cadmium, and so they should not be eaten in very large quantities. One should also avoid fruitbodies from potentially contaminated places, such as roadsides in towns and cities.

•• **Horse Mushroom** (*Agaricus arvensis*) forms medium to large, white fruitbodies which slowly turn yellow where bruised. You typically do not notice the yellowing reaction upon picking, but after an hour in the basket, the pale yellow spots on the cap and stem are evident. Like all royals, it has a well-developed ring with a cogwheel-like structure below. When fruitbodies are damaged, they emit a strong, delicious smell of marzipan, which becomes even more evident when you throw the mushrooms in a hot pan.

Horse Mushroom mostly grows in lawns and fields, and fruits from early summer to autumn.

There are several closely related species with the same culinary properties, such as the more long-stemmed Wood Mushroom (*Agaricus silvicola*), which prefers to grow in woods and forests, and the brown-capped Prince (*Agaricus augustus*) which is found in parks, gardens and open woodland. All are excellent edible fungi.

OTHER SIMILAR FUNGI:

– ††**Destroying Angel** (**deadly poisonous**) always has white gills, see pages 130 and 134.

– **Yellow Stainer** turns deep yellow immediately when scratched at the cap margin and smells of chemicals, see opposite. It can cause stomach upsets.

Cogwheel rings in a young and a mature specimen

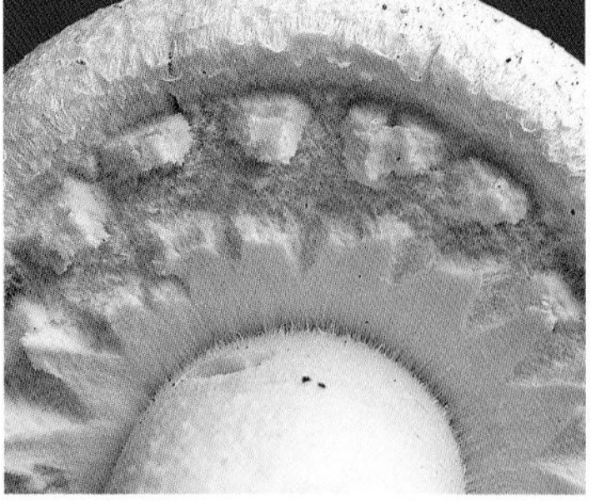

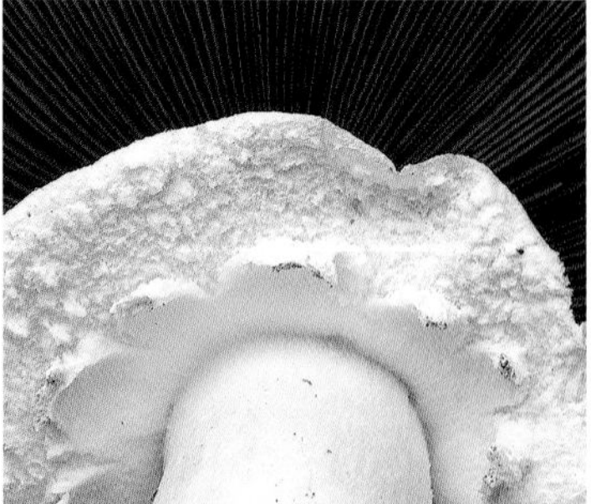

The fairies have danced during the night and summoned the mushrooms

• **Yellow Stainer** (*Agaricus xanthodermus*) forms medium-sized fruitbodies with a slightly angular cap. It can be chalk-white or greyish and has a well-developed ring with a coarse-toothed cogwheel beneath. If you scratch a young and fresh fruitbody at the cap margin, this immediately turns bright yellow. This reaction is usually also seen in the flesh of the stem base. If the fungus is dry, moisten your finger with saliva before rubbing. The smell is unpleasant and chemical (like phenol) and as an extra warning, the smell is amplified if the fruitbodies end up in a hot pan.

Some people can eat the Yellow Stainer without any problems, but in most it causes a stomach upset. It should therefore be avoided.

Yellow Stainers are mostly found in parks and cemeteries (see also the picture on page 16), where they may be quite common during summer and autumn.

***Inkcaps** have free gills that dissolve at maturity and drip away as a black liquid ('ink').*

Many inkcaps have very crowded gills, but because these drip away from the outside in, the spores can still be shot freely into the air from basidia (see page 46) just above the dissolving area. The crowded gills allow the fruitbodies to produce an incredible number of spores.

Inkcaps are decomposers. Most form quite small fruitbodies and are completely uninteresting as edible mushrooms. Only Shaggy Inkcap, with its tall, white fruitbodies, can be used.

Mature Shaggy Inkcap

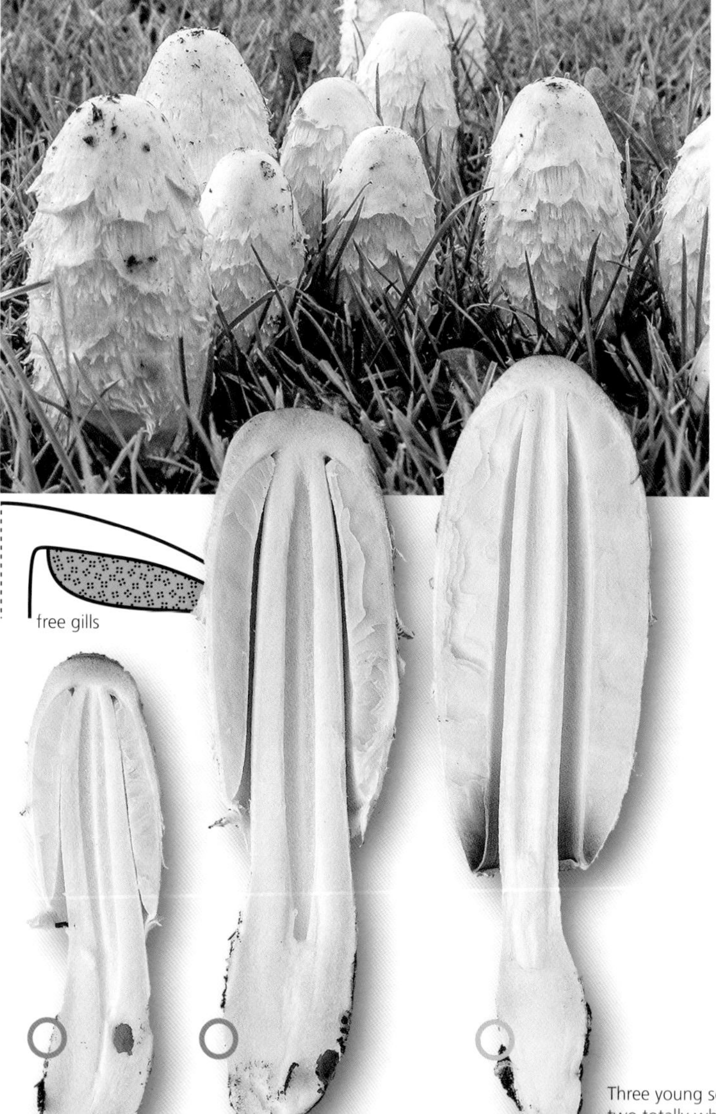

Three young sectioned Shaggy Inkcaps – only the two totally white ones should be eaten

• **Shaggy Inkcap** (*Coprinus comatus*), also known as Lawyer's Wig, is the inkcap with the largest fruitbodies. When mature, they can easily grow to 40 cm tall, and even when very young, they provide a good basketful. The young caps are whitish or light brown and much taller than wide. As soon as they mature, the surface begins to crack into large, curved scales. The stem is white and hollow and has a small, loose ring that easily fall off. The gills are white at first, but are soon coloured dark by maturing spores from the bottom upwards.

OTHER SIMILAR FUNGI:
– other species of Inkcap with large fruitbodies, e.g. **Common Inkcap**, do not have the characteristic white, curved scales. They are not highly poisonous, but Common Inkcap has a strong effect like that of Antabuse, used to treat alcohol dependency; see opposite.

Shaggy Inkcap fruits in September and October. It prefers disturbed places like lawns and stubble fields. It is a fine but very soft-fleshed edible mushroom, which, for example, can add flavour to a mushroom sauce or a pie (see page 38). Use only young fruitbodies that are still white inside, and do not store the catch until tomorrow: it will flow away as ink during the night!

A lawnful of mature and inedible Shaggy Inkcaps

• **Common Inkcap** (*Coprinopsis atramentarius*) is a fairly large inkcap that looks a lot like the Shaggy Inkcap. But where the Shaggy Inkcap cracks into large, white scales, Common Inkcap is smoother with only small, adpressed, brown flakes at the top.

Common on roadsides and in gardens, this species is found in both spring and autumn. It is in principle edible, but contains the substance coprin, which affects the body's ability to break down alcohol. Consumption of alcohol even several days after eating Common Inkcap can cause very unpleasant reactions, exactly like those after treatment with Antabuse. **For safety's sake, one should avoid eating it**.

• **Slimy Spike** (*Gomphidius glutinosus*) has grey-brown caps with a slimy upper side. The slime is the remnants of a universal veil that protects the young fruitbodies from dehydration and predation by insects. The gills are decurrent and become grey as the black spores mature. The black spores are quite easy to see as they get caught in the slimy remnants of veil at the top of the stem. The flesh of the stem base is strikingly yellow.

OTHER SIMILAR FUNGI:
The combination of decurrent gills, black spores, yellow stem base and slimy cap makes this species unmistakable.

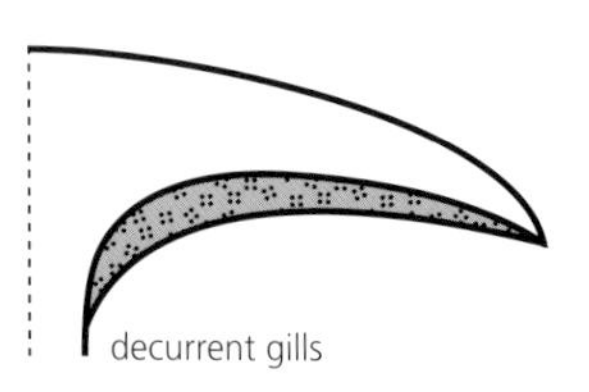

The slimy veil split open to reveal the gills

A sectioned stem shows the yellow flesh

Pull away the slimy cap layer with a knife

Slimy Spike either forms a mycorrhizal association with spruce or is a parasite on the mycelium of other mycorrhizal species. It is very common in spruce forests on poor soil, where it fruits from August to October.

Due to the slimy cap surface, it is not a very inviting edible mushroom. With a little practice one can, however, pull away the slimy layer of the young caps. The same technique can be used in the preparation of slimy boletes from the genus *Suillus*, to which this species is related (surprisingly, given its gills). The greatest value of Slimy Spike is that it is almost impossible to misidentify. Young fruitbodies may be used as a supplement to more tasty species while older specimens should be left in the forest.

•• **The Gypsy** (*Cortinarius caperatus*) forms tall, elegant fruitbodies with a pale greyish-yellow cap, greyish-yellow gills and a whitish stem that is slightly wider at the base. Young caps have small, flaky, pale scales. Around the stem sits a characteristic ring which often has a smal, protruding ridge from the middle. The ring is the remnants of a membranaceous veil that protects the young gills. The spores are rusty brown.

For many years placed in its own genus, *Rozites*, it has been shown that this species belongs among the webcaps (*Cortinarius*), and it is in many ways similar to these (see opposite).

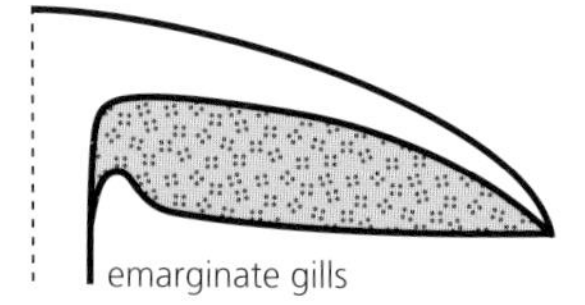

The characteristic ring with a small protruding edge in the middle

The Gypsy is a very fine edible fungus that is collected in many parts of the world. It forms mycorrhizal associations with deciduous and coniferous trees. Unfortunately, it is rare in Britain and Ireland, with most records coming from Scotland. In more mountainous regions of Europe, including the Alps, Norway and Sweden, The Gypsy is much more common.

The Gypsy is weighed by a Bhutanese street seller

OTHER SIMILAR FUNGI:

– **webcaps** (other species of *Cortinarius*) do not have a well-defined ring, but instead show a cobweb-like veil that covers the gills, and is often left in fragments like woolly belts down the stem.

There are some very poisonous webcaps, but these are much more strongly coloured (either dark orange-brown or orange-yellow) and cannot really be confused with The Gypsy (see page 27).

– † **Livid Pinkgill** (*Entoloma sinuatum*) may look rather similar, but it has gills that eventually are coloured yellow-red by the pinkish spores and it has no ring. It grows in nutrient-rich deciduous woodlands **and can cause severe stomach upsets and liver faliure**.

• ***Puffballs*** *form spherical or pear-shaped fruitbodies, which at maturity become filled with brownish spores. When young, they are solid and uniformly white inside and in this state they are edible.*

Puffballs belong to a group called stomach fungi, all of which develop spores inside the fruitbody. These species cannot actively disperse their spores, but instead rely on external forces such as raindrops or boisterous children for spore release.

OTHER SIMILAR FUNGI:

– **earthballs** (*Scleroderma*) resemble puffballs, but they have a tough, outer crust, and the inner spore-mass is black with a marbled pattern. They smell unpleasantly metallic and are slightly poisonous (see opposite).

– young specimens of ***Amanita*** species still wrapped in their universal veil may also look similar. In cross-section, you can see the young, white gills inside the 'ball' (see opposite and page 128). **There are deadly poisonous species in this group**.

Young and old Stump Puffball – the only species on wood

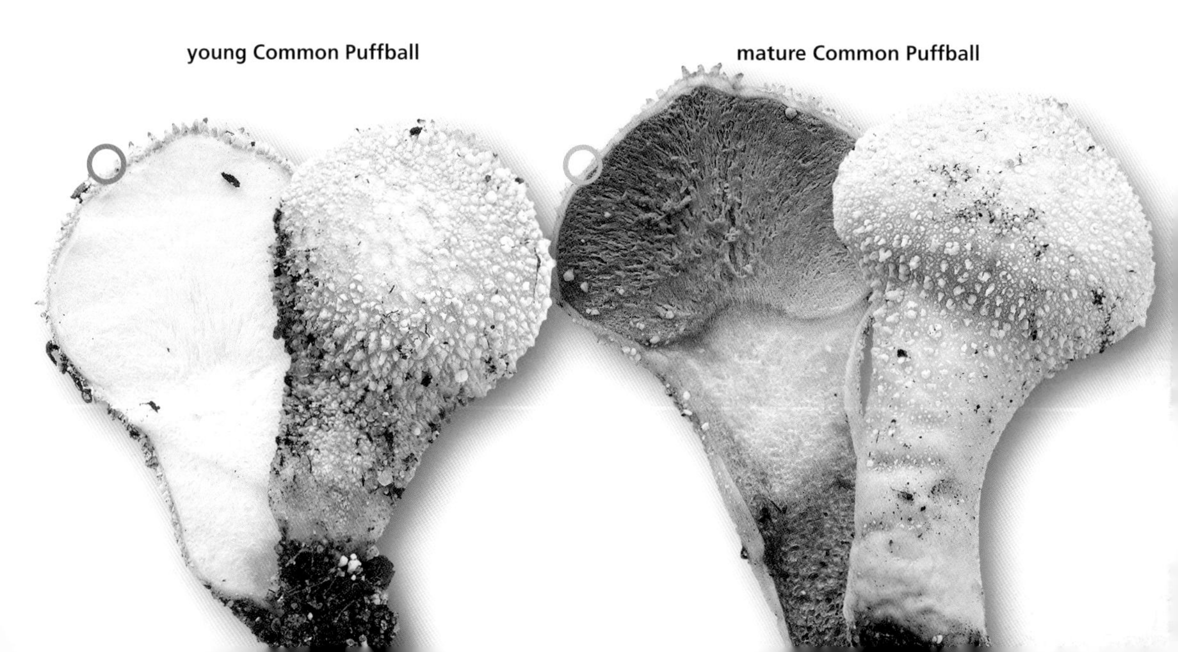

Young and old Mosaic Puffball

Puffballs are decomposers. Some species – e.g. Common Puffball (*Lycoperdon perlatum*) – grow on the ground in forests; some – e.g. Mosaic Puffball (*Lycoperdon utriforme*) – grow in fields; and a single species – Stump Puffball (*Apioperdon pyriforme*) – grows on dead hardwood. All young puffballs are edible, but their culinary value is not high. The best and also the most spectacular is Giant Puffball (see page 148), which can be found on nutrient-rich soil in grazed fields, gardens, hedgerows, scrub and woodland.

Young, edible Giant Puffball – white all the way through. Top right: old dusty winter fruitbodies

• **Giant Puffball** (*Calvatia gigantea*) forms huge, spherical fruitbodies, like footballs or larger, and it can sometimes form impressive fairy rings with many dozens of 'balls'. When young, the fruitbodies are white inside, but with age they fill with billions of brown spores, and as the surface degrades they eventually look like giant, dusty bath sponges.

Smaller puffballs may be confused with young amanitas (see page 147), but luckily the Giant Puffballs are much larger and confusion with amanitas should not be possible.

OTHER SIMILAR FUNGI:

– **Mosaic Puffball** (*Lycoperdon utriforme*), the second largest puffball, rarely exceeds 7–8 cm in diameter (see page 147) and is edible when young.

Fairy rings of Giant Puffball can be very impressive

Giant Puffball is an interesting edible fungus: cut a young solid fruitbody into centimetre-thick slices and coat them in eggs and white breadcrumbs or flour. Then fry in the pan as if it was a plaice. The result has a surprisingly fine consistency, and a mild taste – as one guest exclaimed: "this tastes better than tofu!"

This species typically grows in nutrient-rich soil, such as in gardens and well-fertilized fields, see also the picture on page 14. It is found especially in the southern parts of Britain (more scattered in Ireland) and fruits in August and the beginning of September.

Breaded Giant Puffball

Jelly fungi *are jelly-like or cartilaginous basidiomycetes that mostly grow on wood. Many species form fruitbodies during the winter months.*

There are no poisonous mushrooms among the jelly fungi, but due to their consistency, most species are not considered edible.

In Asian cuisine, two species are used, the White Jelly Fungus (*Tremella fuciformis*) and an Asian species of Jelly Ear.

• **Jelly Ear** (*Auricularia auricula-judae*) forms medium-sized, ear-shaped, brown fruitbodies with a soft, cartilaginous consistency. The upper side is covered with small hairs, while the underside is smooth and somewhat wrinkled.

Jelly Ear prefers to grow on elder but may also be found on many other deciduous trees. It is widespread in most of Britain and Ireland and may be found all year round, even during the depths of winter. As the fruitbodies contain a lot of water they can be difficult to fry. Jelly Ear is, on the other hand, very suitable in oven-baked dishes such as pies and pâtés. Species of Jelly Ear are highly prized in Chinese cuisine and are sold dried as 'Chinese black mushrooms'.

This is the only jelly fungus in our islands that has any significant culinary value.

Other similar fungi:

– **Tripe Fungus** has a much more hairy upper side. It is not poisonous, but neither is it considered an edible fungus, see opposite.

– other **brown jelly fungi** (e.g. from the genera *Tremella*, *Phaeotremella* and *Exidia*) are not ear-shaped, but either top-shaped or brain-like and have a more-or-less smooth surface, see opposite. They are not poisonous, but on the other hand are not worth eating.

• **Tripe Fungus** (*Auricularia mesenterica*) forms protruding, cartilaginous caps with a hairy upper side and rather yellowish-grey colours. The underside of the caps is usually strongly wrinkled or folded – much more pronounced than seen in the Jelly Ear.

It is not considered an edible fungus, but is not poisonous either. It is predominantly found on elm.

• **Leafy Brain** (*Phaeotremella foliacea*) is rather similar to Jelly Ear, but its fruitbodies are even more jelly-like and are folded in a more brain-like manner. It is also not hairy.

It actually constitutes a complex of closely related, very similar species, with fruitbodies on both coniferous and deciduous trees. Here the fungus lives as a parasite upon the mycelium of various crust fungi, like Bleeding Conifer Crust (*Stereum sanguinolentum*) and Bleeding Broadleaf Crust (*Stereum rugosum*), which are decomposing the trees. In the picture you can see the flat, light brown fruitbody of Bleeding Broadleaf Crust below the Leafy Brain.

Leafy Brain is not considered to be an edible mushroom, presumably because of the jelly-like consistency and lack of flavour. It is found scattered in late autumn and winter. There are many similar species of jelly fungi.

The symbols used in this book

- •• excellent after proper cooking
- • acceptable to eat after proper cooking
- • worthless
- † poisonous
- †† deadly poisonous

edible after proper cooking | worthless | poisonous

Index notes – *Italic page numbers* indicate images that are not part of the main species accounts.

Published by Princeton University Press
41 William Street, Princeton, NJ 08540, USA
99 Banbury Road, Oxford OX2 6JX, UK
press.princeton.edu

Layout and prepress: Jens H. Petersen.
Editor: Axel Kielland.
Printing, paper and binding: GPS Group
Printed in Bosnia and Herzegovina

ISBN 978-0-691-24519-5
ISBN (ebook) 978-0-691-24536-2

British Library Cataloging-in-Publication data is available

10 9 8 7 6 5 4 3 2 1

Photographs
All photographs by the author except:
Benny Christensen: 106 (top).
Flemming Rune: 36 (bottom).
Jan Vesterholt: 81 (top), 85 (middle), 93 (top).
Jens Maarbjerg: 5 (second row, left), 30 (bottom), 38 (bottom), 39 (two upper), 59 (bottom), 64 (bottom), 66 (bottom centre).
Morten Christensen: 17.
Thomas Læssøe 58 (top), 75 (bottom left), 86 (middle, left), 91 (top), 93 (top), 104 (middle), 125 (top left).
Tobias Frøslev: 27 (bottom centre).

Thanks to Lise Samsø for the yarn samples page 40-41, and to Thomas Læssøe for reading and commenting on the Danish manuscript.

Jacket images: Various fungi by Jens H. Petersen; background image: iStock

Cover design by Simran Rohira

Learn more about fungi at **The British Mycological Society** britmycolsoc.org.uk/
or consult the two volume book set ***Fungi of Temperate Europe***
or ***The Kingdom of Fungi***

The fungi in this book arranged by where they grow

fruitbodies on the ground in woodland

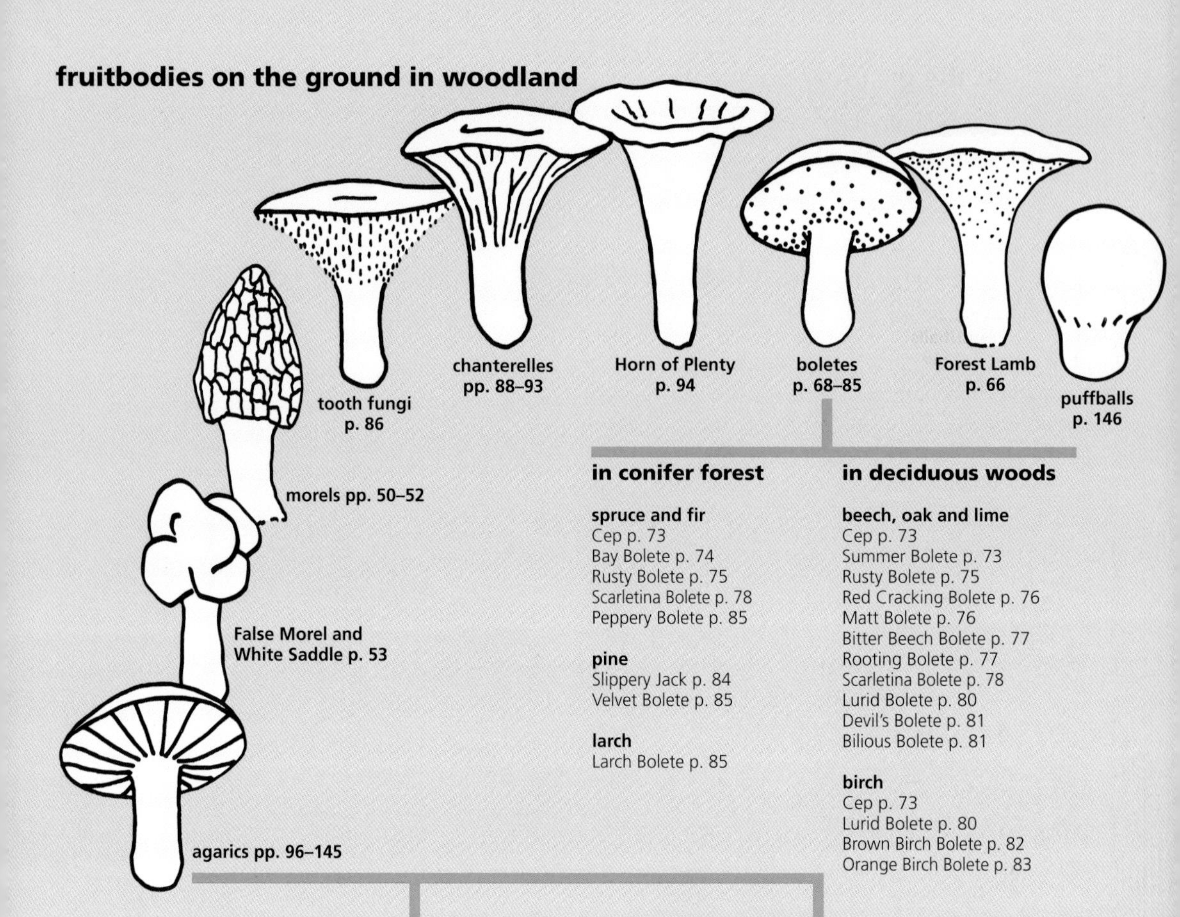

in conifer forest

spruce and fir
Cep p. 73
Bay Bolete p. 74
Rusty Bolete p. 75
Scarletina Bolete p. 78
Peppery Bolete p. 85

pine
Slippery Jack p. 84
Velvet Bolete p. 85

larch
Larch Bolete p. 85

in deciduous woods

beech, oak and lime
Cep p. 73
Summer Bolete p. 73
Rusty Bolete p. 75
Red Cracking Bolete p. 76
Matt Bolete p. 76
Bitter Beech Bolete p. 77
Rooting Bolete p. 77
Scarletina Bolete p. 78
Lurid Bolete p. 80
Devil's Bolete p. 81
Bilious Bolete p. 81

birch
Cep p. 73
Lurid Bolete p. 80
Brown Birch Bolete p. 82
Orange Birch Bolete p. 83

in deciduous woodlands

with beech

The Flirt p. 100
The Sickener p. 99
Olive Brittlegill p. 101
Ochre Brittlegill p. 103
Geranium Brittlegill p. 103
Greencracked Brittlegill p. 105
Charcoal Burner p. 105
Fishy Milkcap p. 108
Birch Milkcap p. 109
Panthercap p. 127
Deathcap p. 129
Destroying Angel p. 130
Fly Agaric p. 131
Scaly Wood Mushroom p. 136
Horse Mushroom p. 138
The Gypsy p. 144

with birch

The Flirt p. 100
Yellow Swamp Brittlegill p. 102
Green Brittlegill p. 104
Fenugreek Milkcap p. 112
Ugly Milkcap p. 113
Woolly Milkcap p. 113
Bearded Milkcap p. 113
Fly Agaric p. 131

with oak

Ochre Brittlegill p. 103
Geranium Brittlegill p. 103
Oakbug Milkcap p. 109
Birch Milkcap p. 109
Deathcap p. 129
Destroying Angel p. 130

in coniferous forest

with pine

Crab Brittlegill p. 97
Hintapink p. 98
The Sickener p. 99
Darkening Brittlegill p. 101
Copper Brittlegill p. 102
Saffron Milkcap p. 110
Fenugreek Milkcap p. 112
Fly Agaric p. 131

with spruce

Crab Brittlegill p. 97
The Sickener p. 99
Hintapink p. 100
Ochre Brittlegill p. 103
Birch Milkcap p. 109
Rufous Milkcap p. 109
False Saffron Milkcap p. 111
Fenugreek Milkcap p. 112
Ugly Milkcap p. 113
Destroying Angel p. 130
Fly Agaric p. 131
Wood Blewit p. 133
Scaly Wood Mushroom p. 136
Horse Mushroom p. 138
Slimy Spike p. 142

with fir

Crab Brittlegill p. 97
Hintapink p. 98
Ochre Brittlegill p. 103
Ugly Milkcap p. 113
Fly Agaric p. 131